OPPORTUNITIES IN NEUROSCIENCE FOR FUTURE ARMY APPLICATIONS

Committee on Opportunities in Neuroscience for Future Army Applications

Board on Army Science and Technology

Division on Engineering and Physical Sciences

NATIONAL RESEARCH COUNCIL
OF THE NATIONAL ACADEMIES

THE NATIONAL ACADEMIES PRESS
Washington, D.C.
www.nap.edu

THE NATIONAL ACADEMIES PRESS **500 Fifth Street, N.W.** **Washington, DC 20001**

NOTICE: The project that is the subject of this report was approved by the Governing Board of the National Research Council, whose members are drawn from the councils of the National Academy of Sciences, the National Academy of Engineering, and the Institute of Medicine. The members of the committee responsible for the report were chosen for their special competences and with regard for appropriate balance.

This study was supported by Contract No. W911NF-07-C-0117 between the National Academy of Sciences and the Department of Defense. Any opinions, findings, conclusions, or recommendations expressed in this publication are those of the author(s) and do not necessarily reflect the views of the organizations or agencies that provided support for the project.

International Standard Book Number-13: 978-0-309-12740-0
International Standard Book Number-10: 0-309-12740-8
Library of Congress Control Number: 2009927221

Limited copies of this report are available from

Board on Army Science and Technology
National Research Council
500 Fifth Street, N.W., Room 940
Washington, DC 20001
(202) 334-3118

Additional copies are available from

The National Academies Press
500 Fifth Street, N.W.
Lockbox 285
Washington, DC 20055
(800) 624-6242 or
(202) 334-3313
(in the Washington metropolitan area)
Internet, http://www.nap.edu

Cover: *Neuronal pathways in BrainBow mice. Neurons in the hippocampus, a brain area involved in memory, are labeled in different colors, with their neuronal outgoing projections pointing to the left. This is the first time so many different neurons have been separately visualized on such a large scale. Courtesy of Jean Livet, Joshua R. Sanes, and Jeff W. Lichtman, Harvard University, 2008.*

Printed in the United States of America

THE NATIONAL ACADEMIES

Advisers to the Nation on Science, Engineering, and Medicine

The **National Academy of Sciences** is a private, nonprofit, self-perpetuating society of distinguished scholars engaged in scientific and engineering research, dedicated to the furtherance of science and technology and to their use for the general welfare. Upon the authority of the charter granted to it by the Congress in 1863, the Academy has a mandate that requires it to advise the federal government on scientific and technical matters. Dr. Ralph J. Cicerone is president of the National Academy of Sciences.

The **National Academy of Engineering** was established in 1964, under the charter of the National Academy of Sciences, as a parallel organization of outstanding engineers. It is autonomous in its administration and in the selection of its members, sharing with the National Academy of Sciences the responsibility for advising the federal government. The National Academy of Engineering also sponsors engineering programs aimed at meeting national needs, encourages education and research, and recognizes the superior achievements of engineers. Dr. Charles M. Vest is president of the National Academy of Engineering.

The **Institute of Medicine** was established in 1970 by the National Academy of Sciences to secure the services of eminent members of appropriate professions in the examination of policy matters pertaining to the health of the public. The Institute acts under the responsibility given to the National Academy of Sciences by its congressional charter to be an adviser to the federal government and, upon its own initiative, to identify issues of medical care, research, and education. Dr. Harvey V. Fineberg is president of the Institute of Medicine.

The **National Research Council** was organized by the National Academy of Sciences in 1916 to associate the broad community of science and technology with the Academy's purposes of furthering knowledge and advising the federal government. Functioning in accordance with general policies determined by the Academy, the Council has become the principal operating agency of both the National Academy of Sciences and the National Academy of Engineering in providing services to the government, the public, and the scientific and engineering communities. The Council is administered jointly by both Academies and the Institute of Medicine. Dr. Ralph J. Cicerone and Dr. Charles M. Vest are chair and vice chair, respectively, of the National Research Council.

www.national-academies.org

COMMITTEE ON OPPORTUNITIES IN NEUROSCIENCE FOR FUTURE ARMY APPLICATIONS

FLOYD E. BLOOM, *Chair*, Professor Emeritus, The Scripps Research Institute
RICHARD A. ANDERSON, Professor of Neuroscience, California Institute of Technology
RONALD R. BLANCK, Vice Chairman, Martin, Blanck, & Associates
EMERY N. BROWN, Professor of Anesthesia, Massachusetts General Hospital
JOSEPH T. COYLE, Eben S. Draper Professor of Psychiatry and Neuroscience, Harvard University
MARY "MISSY" CUMMINGS, Boeing Assistant Professor, Director of Humans and Automation Lab, Massachusetts Institute of Technology
J. MARK DAVIS, Professor, Director of Psychoneuroimmunology, Division of Applied Physiology, University of South Carolina
MICHAEL S. GAZZANIGA, Director, The Sage Center for the Study of the Mind, University of California at Santa Barbara
RICHARD J. GENIK, II, Director, Emergent Technology Research Division, Wayne State University
PAUL W. GLIMCHER, Professor of Neural Sciences, Economics, and Psychology, New York University
PETER A. HANCOCK, Provost, Distinguished Research Professor, University of Central Florida
STEVEN KORNGUTH, Director, Center for Strategic and Innovative Technologies, Professor of Pharmacy, University of Texas
MARTIN P. PAULUS, Professor in Residence, Department of Psychology, University of California at San Diego
JUDITH L. SWAIN, Executive Director, Singapore Institute of Clinical Sciences; Lien Ying Chow Professor of Medicine, National University of Singapore; Adjunct Professor of Medicine, University of California at San Diego
PAUL J. ZAK, Director, Center for Neuroscience Studies, Professor of Economics and Clinical Professor of Neurology, Claremont Graduate University

Staff

ROBERT J. LOVE, Study Director
NIA D. JOHNSON, Senior Research Associate (through July 2008)
JAMES C. MYSKA, Senior Research Associate
ANGELA L. MARTIN, Senior Program Assistant

BOARD ON ARMY SCIENCE AND TECHNOLOGY

Staff

Preface

The study of the human brain, its capacities, and its diseases remains one of the greatest scientific and philosophical challenges ever undertaken. That said, compiling this report to consider aspects of what has been learned and is being learned about the brain that can be useful to the operations of the U.S. Army has also proven to be extremely challenging. First, the field of neuroscience is so dynamic that significant new findings are being announced almost daily. The ability to define the systems of neurons that are activated as a human learns, practices, and performs mental tasks has allowed impressive integration of the constantly accruing understanding of how neurons use their genes to create proteins that enable them to function together. The ability to correlate brain images with function and behavior has already translated into accelerated programs to develop yet more advanced imaging tools and techniques. I believe, as many others do, that despite the almost constant growth of neuroscience over the past four decades, the future of neuroscience applications will grow at a rate that has not been seen since the birth of microprocessor-based personal computers. It was against this backdrop of a rapidly growing science, and the even more rapidly changing translation of that science into useful applications, that the committee addressed the issues in this report.

Second, along with the growing number of scientists engaged in neuroscience, the scope of neuroscience is also expanding at an ever-accelerating rate—so fast, in fact, that scientists and engineers have difficulty reconciling their perceptions of what is and isn't included. Several new subdisciplines have been created in the past few years, easily identified by the "neuro-" prefix, linking neuroscience knowledge with valuable applications and technologies both medical and nonmedical. Which of them have proven to be substantive enough to be declared authentic components of applied neuroscience and to be deemed contributors to that future? Yet a third challenge has been to prepare a report that could satisfy and be understood by an audience of both generalists and specialists, as well as by those in the Army who must make the hard decisions on which science and technology to pursue.

The committee examined the basis of new neuroscience-based technologies and the likelihood that they could one day have an impact on Army capabilities. I believe the report provides both a valuable snapshot of the nature of neuroscience today and a well-formulated conceptualization of how its growing number of facets could affect the Army. Although the science is a moving target, actions could be taken now to track the progression of new concepts that will lead to developments with high potential for Army use.

I would like to thank the committee for its hard work in interviewing numerous experts, assessing the pertinent issues, and developing recommendations to address the many demands of its statement of task (see page 9). The committee, in turn, is grateful to the many Army personnel engaged in related research and technology developments for the useful information they provided. We also greatly appreciate the support and assistance of the National Research Council (NRC) staff, which ably assisted the committee in its fact-finding activities and in the production of this report. In particular, I thank Robert Love and his staff, who successfully organized several major meetings in multiple locations. They also maintained a secure central Web forum, where our guests' messages accumulated, along with remote interviews with people unable to attend our meetings, and through which we wrote, shared our expertises, and developed the consensus for the report we present here. In addition, we also specifically recognize the essential role played by consulting technical writer Robert Katt as he helped us communicate to our intended audiences within the Army the richness of the data and the often subtle nuances of how those data could be used by the Army for its current and future operations.

The study was conducted under the auspices of the NRC Board on Army Science and Technology (BAST), which was established as a unit of the NRC in 1982 at the request of the U.S. Army. The BAST brings broad military, industrial, and academic scientific, engineering, and management expertise to bear on technical challenges of importance to senior Army leaders. The Board is not a study committee; rather, it discusses potential study topics and tasks; ensures project planning; suggests potential experts to serve as committee members or reviewers; and convenes meetings to examine strategic issues for its sponsor, the Assistant Secretary of the Army (Acquisition, Logistics, and Technology).

Although the Board members are listed on page *vi* of this report, they were not, with the exception of any Board members nominated and appointed to serve as formal members of the study committee, asked to endorse the committee's conclusions or recommendations, nor did they review final drafts of the report before its release.

Floyd Bloom, *Chair*
Committee on Opportunities in Neuroscience for Future Army Applications

Acknowledgments

This report has been reviewed in draft form by individuals chosen for their diverse perspectives and technical expertise, in accordance with procedures approved by the National Research Council's Report Review Committee. The purpose of this independent review is to provide candid and critical comments that will assist the institution in making its published report as sound as possible and to ensure that the report meets institutional standards for objectivity, evidence, and responsiveness to the study charge. The review comments and draft manuscript remain confidential to protect the integrity of the deliberative process. We wish to thank the following individuals for their review of this report:

Huda Akil, University of Michigan,
Michael Bennett (NAS), Albert Einstein College of Medicine,
John Cacioppo, University of Chicago,
David Dinges, University of Pennsylvania,
Stephen Drew, Science Partners,
Christopher Green, Wayne State University,
Edward G. Jones (NAS), University of California, Davis,
Michael M. Merzenich (NAS), University of California, San Francisco,
Gregory J. Moore, Pennsylvania State University, and
Daniel Schacter, Harvard University.

Although the reviewers listed above have provided many constructive comments and suggestions, they were not asked to endorse the conclusions or recommendations, nor did they see the final draft of the report before its release. The review of this report was overseen by Dennis W. Choi of Emory University and Richard M. Shiffrin of Indiana University. Appointed by the National Research Council, they were responsible for making certain that an independent examination of this report was carried out in accordance with institutional procedures and that all review comments were carefully considered. Responsibility for the final content of this report rests entirely with the authoring committee and the institution.

Contents

Tables, Figures, and Boxes

TABLES

FIGURES

BOXES

Acronyms and Abbreviations

AFAST	Alternative Flight Aptitude Selection Test
AFQT	Armed Forces Qualification Test
AIM	assessment of individual motivation
ANAM	Automated Neuropsychological Assessment Metrics
AR	augmented reality
ARI	Army Research Institute
ASVAB	Armed Services Vocational Aptitude Battery
ATP	adenosine triphosphate
AugCog	augmented cognition
BAST	Board on Army Science and Technology
BCAA	branched-chain amino acid
BMI	brain–machine interface
BOLD	blood oxygen level–dependent
CNS	central nervous system
CRT	cathode ray tube
CSF	cerebral spinal fluid
C2V	command and control vehicle
DARPA	Defense Advanced Research Projects Agency
DOT	diffuse optical tomography
DSI	diffusion spectrum imaging
DTI	diffusion tensor imaging
EEG	electroencephalography
ERP	event-related potential
FDA	Food and Drug Administration
fMRI	functional magnetic resonance imaging
GSR	galvanic skin response
HMI	human–machine interface
HPA	hypothalamic-pituitary-adrenal
IED	improvised explosive device
LCD	liquid crystal display
lfp	local field potential
MEG	magnetoencephalography
MRI	magnetic resonance imaging
NASA	National Aeronautics and Space Administration
NIRS	near-infrared spectroscopy
NMDA	N-methyl-D-aspartic acid
NonREM	nonrapid eye movement
NRC	National Research Council
NSBRI	National Space and Biomedical Research Institute
OEM	original equipment manufacturer
OT	oxytocin
PET	positron emission tomography
PFC	prefrontal cortex
PTSD	post-traumatic stress disorder
PVT	Psychomotor Vigilance Test
R&D	research and development
REM	rapid eye movement
RPD	recognition-primed decision
SAT	standardized assessment test
SIFT	Selection Instrument for Flight Training
T	tesla
TBI	traumatic brain injury
TMS	transcranial magnetic stimulation
ToM	theory of mind
TPJ	temporoparietal junction
VR	virtual reality

Summary

Emerging neuroscience opportunities have great potential to improve soldier performance and enable the development of technologies to increase the effectiveness of soldiers on the battlefield. Advances in research and investments by the broader science and medical community promise new insights for future military applications. These include traditional areas of interest to the Army, such as learning, decision making, and performance under stress, as well as new areas, such as cognitive fitness, brain–computer interfaces, and biological markers of neural states.

Advances in research-enabling technologies, such as functional magnetic resonance imaging (fMRI) and computational neuroscience, have resulted in instrumentation and techniques that can better assess the neural basis of cognition and allow the visualization of brain processes. These advances have the potential to provide new measures of training and learning for soldiers while also shedding fresh light on the traditional approaches to behavioral science used by the Army. Most current Army neuroscience research is conducted with little regard given to its longer-term potential for military operations. The report discusses a spectrum of ongoing efforts, with an emphasis on nonmedical applications and on current research that is likely to lead to insights and opportunities for possible military application.

STUDY APPROACH

The Assistant Secretary of the Army (Acquisition, Logistics, and Technology) (ASAALT) asked the National Research Council to conduct a study of neuroscience in terms of its potential to support military applications. Chapter 1 discusses the statement of task and explains how the report responds to each of the tasks.

Members of the Committee on Opportunities in Neuroscience for Future Army Applications, set up in response to the ASAALT request, had expertise not only in traditional and emerging subdisciplines of neuroscience but also in research and development (R&D), in military operations and medicine, and in training specialties such as memory and learning, assessment, decision making, prediction, and reading intentionality. The short biographies of the committee members, given in Appendix A, include their specialties.

Early briefings on the scope of Army research in neuroscience provided the basis for dividing the committee into data-gathering teams. The committee's meetings and data-gathering activities are described in Appendix B. The various streams of information were brought together and a consensus was reached on the conclusions and recommendations of the report. The information gathered by the teams is organized into chapters on training and learning, optimizing decision making, sustaining soldier performance, improving cognitive and behavioral performance, neuroscience technology opportunities, and long-term trends.

The committee was tasked to identify research and technology development opportunities and to recommend those worthy of investment in the near, medium, and far terms. High-payoff research opportunities are provided as Recommendations 1 through 15. The topics considered to be technology development opportunities were judged high priority (Table S-1), priority (Table S-2), and worthy of monitoring for possible future implementation (Table S-3). The committee considered all topics in Tables S-1 and S-2 worthy of immediate investment. Prioritization of the opportunities within the "high-priority" group and the "priority" group is dependent on the relative importance to the Army of the particular applications served by the topics.

The committee's consensus was that in the near term the Army would benefit primarily from advances in cognitive neuroscience—education, assessment, and training, as described in Chapters 3-6. Advances in molecular and cellular neuroscience were not judged likely to have as much impact on Army operations, and the chance that advances in systems neuroscience would have an impact was quite remote. In addition, the committee considered that noninvasive, technology-based research would be the most likely to lead to discernible Army applications in the time frame

TABLE S-1 High-Priority Opportunities for Army Investment in Neuroscience Technologies (Recommendation 14)

				Current Investment (L, M, or H)	
Technology Opportunity	ME	RE	Time Frame[a]	Commercial	Academic
Field-deployable biomarkers of neural state	x	x	Ongoing	L	M
In-helmet EEG for brain–machine interface	x	x	Medium term	M	L
Signal processing and multimodal data fusion, including imaging modalities such as MRI, fMRI, DTI, DSI, PET, and MEG and physiological measures such as heartbeat, interbeat intervals, GSR, optical computer recognition, eye tracking, and pupilometry.	x	x	Ongoing	M	H
Soldier models and biomarkers for sleep		x	Ongoing	M	M
Vertical fMRI		x	Medium term	L	L
Fatigue prediction models	x		Medium term	L	M
Behavioral measures of fatigue	x		Medium term	M	L
Prospective biomarkers for predictive measures of soldier response to environmental stress, including hypoxic and thermal challenges	x	x	Medium term	L	L
NIRS/DOT	x	x	Medium term	L	L
Biomedical standards and models for head impact protection, including torso protection from blast	x	x	Medium term	M	M
Threat assessment augmentation	x		Medium term	M	M
fMRI paradigms of military interest		x	Ongoing	L	M

NOTE: ME, mission-enabling; RE, research-enabling; L/M/H, low, medium, or high; EEG, electroencephalography; MRI, magnetic resonance imaging; fMRI, functional magnetic resonance imaging; DTI, diffuse tensor imaging; DSI, diffusion spectrum imaging; PET, positron emission tomography; MEG, magnetoencephalography; NIRS, near-infrared spectroscopy; DOT, diffuse optical tomography; GSR, galvanic skin response.

[a]In this column, "medium term" means between 5 and 10 years and "ongoing" means that results will be available within 5 years but continuing investment is recommended to stay at the forefront of the technology.

SOURCE: Committee-generated.

TABLE S-2 Priority Opportunities for Army Investment in Neuroscience Technologies (Recommendation 15)

				Current Investment (L, M, or H)	
Technology Opportunity	ME	RE	Time Frame[a]	Commercial	Academic
Haptic feedback with VR	x		Medium term	H	L
Augmented reality (virtual overlay onto real world)	x	x	Medium term	H	H
In-helmet EEG for cognitive state detection and threat assessment	x	x	Medium term	L	M
Information workload management	x		Far term	L	M
Time-locked, in-magnet VR and monitoring for fMRI		x	Medium term	L	M
Immersive, in-magnet VR		x	Near term	L	M
EEG physiology	x	x	Far term	L	H
Uses of TMS for attention enhancement		x	Medium term	L	M
In-vehicle TMS deployment	x		Far term	L	L
Heartbeat variability	x	x	Near and medium term	L	H
Galvanic skin response	x	x	Near and medium term	H	L

NOTE: ME, mission-enabling; RE, research-enabling; L/M/H, low, medium, or high; VR, virtual reality; TMS, transcranial magnetic stimulation.

[a]In this column, "near term" means within 5 years, "medium term" means between 5 and 10 years, and "far term" means 10-20 years.

SOURCE: Committee-generated.

of the study (5-20 years); however, R&D in a few areas of neuroscience might well have enough potential that the Army should assemble a group of experts charged with monitoring progress on multiple fronts.

In addition to the 15 recommendations that respond to points in the statement of task, the committee offers two overarching recommendations, 16 and 17, which it believes are essential if the Army is to engage the opportunities in neuroscience effectively. Recommendation 16 concerns the establishment of a mechanism to monitor advances in a wide range of neuroscience disciplines in order to stay abreast of new developments and select for further pursuit those most promising for Army applications. Recommendation 17 encourages the Army to examine how to use to its advantage

TABLE S-3 Possible Future Opportunities (Neuroscience Areas Worthy of Monitoring for Future Army Investment)

Technology Opportunity	ME	RE	Time Frame[a]	Current Investment (L, M, or H) Commercial	Academic
Brain–computer interface system (direct)	x		Far term	H	H
Imaging cognition		x	Far term	L	H
Neuropharmacological technology		x	Far term	M	M
Advanced fMRI data collection		x	Medium term	M	M
Averaging methodology for fMRI		x	Medium term	L	M
Brain database aggregation		x	Far term	M	M
Default mode networks	x	x	Medium term	L	H
Inverse MRI		x	Medium term	L	M
Low-field MRI	x	x	Far term	L	M
Uses of TMS for brain network inhibition		x	Far term	L	M
Safety of multiple exposures to TMS		x	Medium term	M	M
In-helmet TMS deployment	x		Far term	L	L
Connectomics		x	Far term	L	M
Atomic magnetometers	x	x	Far term	M	M

NOTE: ME, mission-enabling; RE, research-enabling; L/M/H, low, medium, or high; fMRI, functional magnetic resonance imaging; MRI, magnetic resonance imaging; and TMS, transcranial magnetic stimulation.

[a]In this column, "medium term" means between 5 and 10 years and "far term" means 10-20 years.

SOURCE: Committee-generated.

the insights on individual variability and the human dimension that are emerging from neuroscience.

OPPORTUNITIES IN ARMY APPLICATIONS AREAS

Opportunities exist for the Army to benefit from research in neuroscience by applying and leveraging the results of work by others (including academic research and R&D by other federal agencies and the commercial sector) or by making selective investments in Army-specific problems and applications.

Training and Learning

Neuroscience can extend and improve the Army's traditional behavioral science approaches to both training and learning. For example, neuroscience offers new ways to assess how well current training paradigms and accepted assumptions about learning achieve their objectives. Neuropsychological indicators can help to assess how well an individual trainee has assimilated mission-critical knowledge and skills. These assessment tools also will allow the Army to assess individual variability and tailor training regimens to the individual trainee.

Recommendation 1. The Army should adjust its research capabilities to take advantage of the current and emerging advances in neuroscience to augment, evaluate, and extend its approaches to training and learning. Indicators of knowledge and skill acquisition based in neuroscience should be incorporated into the methods of testing for training success. In particular, these indicators should be employed in identifying individual variability in learning and tailoring training regimens to optimize individual learning.

The Army currently relies heavily on broad, general indicators of aptitude to predict training effectiveness and individual success rates. The importance of predicting success rates of soldiers before assigning them to given tasks increases with the cost of training for the task and with the consequences of not performing the task well. In comparison with the indicators that have been developed for assessing how well skills or knowledge have been acquired, neurological predictors of soldier performance need much R&D before they will be ready for Army applications.

Recommendation 2. The Army should investigate neuropsychological testing of candidates for a training course that is already established as a requirement to enter a high-value field. In this way the Army can determine whether an assignment-specific neuropsychological profile can be developed that has sufficiently high predictive value to use in conjunction with established criteria for the assignment. If results for this investigation are positive, the Army should investigate development of assignment-specific profiles for additional assignments.

Optimizing Decision Making

Human decision making is predictably inefficient and often suboptimal, especially when the decisions require assessments of risk and are made under pressure. Individuals also differ in their approach to making decisions. For example, some individuals are more impulsive, while others are more deliberate and less tolerant of risk. These differences do not mean that risk-tolerant individuals are necessarily better or worse decision makers than risk-averse individuals. From an institutional (Army) point of view, different decision-making styles can suit different individuals for different tasks. That is to say, a given task may require or be better performed by an individual with a particular decision-making style. With research, neuroscience tools may become capable of discerning neural correlates for differences in decision-making style.

Recommendation 3. The Army should expand existing research in behavioral and social sciences to include neuroscience aimed at developing training and assessment tools for decision makers at all levels in the Army.

Sustaining Soldier Performance

The committee reviewed neuroscience applications related to understanding, monitoring, and preventing or treating deficits in soldier performance. These deficits may affect performance during a single extended operation or over much longer time frames. The report considered prevention interventions relevant to acute deficits noticeable immediately within the time frame of a day or days as well as longer-term deficits such as post-traumatic stress disorder (PTSD) and other chronic effects of brain trauma on the central nervous system (CNS).

Individual Variability of Soldiers

Conventional Army operations emphasize common levels of operational readiness and performance among individuals rather than individual variability as the basis for unit effectiveness. Nevertheless, individual soldiers do vary, not only in their baseline optimum performance (i.e., performance not degraded by sustained stressors) but also in their response to stressors that frequently lead to less-than-optimal performance (performance deficits). The Army acknowledges and uses individual variability to its advantage to achieve desired objectives for some high-value assignments that are very dependent on exceptionally high-performing individuals, such as in Special Operations.

An important lesson from neuroscience is that the ability to sustain and improve performance can be increased by identifying differences in individual soldiers and using individual variability to gauge optimum performance baselines, responses to performance-degrading stressors, and responses to countermeasures to such stressors.

Recommendation 4. To increase unit performance across the full spectrum of operations, the Army should expand its capacity to identify and make use of the individual variability of its soldiers. The Army should undertake R&D and review its training and doctrine to take best advantage of variations in the neural bases of behavior that contribute to performance. In particular, it should seek to understand—and use more widely—individual variability in (1) baseline optimal performance, (2) responses to stressors likely to degrade optimal performance, and (3) responses to countermeasures intended to overcome performance deficits or to interventions intended to enhance performance above an individual's baseline.

Countermeasures to Environmental Stressors

The degradation of performance during sustained periods of physical or mental stress results from both peripheral system (e.g., muscle and cardiovascular) and CNS factors, which are inextricably linked. However, we lack sufficient fundamental understanding of how these factors interact and how they are influenced by the range of environmental stressors to which soldiers engaged in sustained operations are exposed. For example, physical and mental fatigue, commonly assumed to result in less-than-optimal performance, are neither well enough defined nor sufficiently well understood to provide a scientific basis for developing effective countermeasures to both the CNS and peripheral components.

Current nutritional countermeasures to fatigue are based primarily on maintaining cardiovascular and muscle function, but they fall short of addressing the important role of nutrition in brain functioning affected by fatigue. One reason for this shortfall is insufficient understanding of the CNS factors that are linked to the stress-induced degradation of performance, including those deficits commonly attributed to physical and/or mental fatigue.

Recommendation 5. The Army should increase both the pace of and its emphasis on research designed to understand the neural bases of performance degradation under stress, including but not limited to deficits commonly attributed to fatigue, and the interaction of peripheral and CNS factors in responses to stressors. It should apply the results of this research to develop and improve countermeasures such as nutritional supplements and management of sleep/wake and rest/wakefulness cycles.

Sleep is an active process that plays a fundamental role in cognitive functions such as consolidating memory and promoting synaptic plasticity. Prolonged sleep deprivation interferes with these functions and can thus adversely affect performance.

Recommendation 6. Since many abilities affected by sleep deprivation—vigilance, memory, and perceptual discrimination, for example—are increasingly important elements of soldier performance, the Army should increase its efforts to collaborate with the lead laboratories involved in physiological and molecular research on sleep.

Pharmaceutical Countermeasures to Performance Degradation

Advances in neuroscience are enabling the pharmaceutical industry to develop drugs that act on novel targets to affect mood, motivation, memory, and executive function.

Recommendation 7. The Army should establish relationships with the pharmaceutical industry, the National Institutes of Health, and academic laboratories to keep abreast of advances in neuropharmacology, cellular and molecular neurobiology, and neural development and to identify new drugs that have the potential to sustain or enhance performance in military-unique circumstances. However, caution must be exercised to ensure that the benefits outweigh any unforeseen or delayed side effects.

Among the neuropharmaceuticals approved by the Food and Drug Administration for specific medical indications, a number have potential off-label uses in sustaining or optimizing performance. However, any compound, natural or synthetic, that acts on the CNS must be assumed, until proven otherwise, to affect multiple neural systems. It is therefore essential that specificity of action be demonstrated. Second, the risks of unforeseen or delayed side effects must be considered, particularly before a neuropharmaceutical is widely administered for sustaining or enhancing performance in mission-critical tasks without specific medical indication to justify its use.

Recommendation 8. Before the Army attempts to employ neuropharmaceuticals for general sustainment or enhancement of soldier performance, the Army should undertake medically informed evidence-based risk-benefit analyses, including performance and clinical measures to assess overall effects, to ensure that the expected benefits of such medication outweigh the risks of negative side effects or delayed effects.

Use of new pharmacological agents to restore function, mitigate pain or other responses to trauma, or facilitate recovery from injury or trauma will be a key application for new neuroscience technology in the near to medium term. Highly specific brain receptor targets have been identified for a number of agents, and the effectiveness of these agents will be greatly enhanced by technologies that target delivery of the pharmacological agent to a specific site. The use of targeted drug delivery to enhance performance, such as situational awareness, is technically feasible, but such uses may be proscribed by societal and ethical norms.

Recommendation 9. The Army should support research on novel mechanisms for noninvasive, targeted delivery of pharmacological agents to the brain and nervous system in the course of medical interventions to mitigate the adverse effects of physical injury to the brain or another portion of the nervous system. In the near to medium term, this research should focus on restoring a performance deficit to baseline function rather than enhancing performance beyond that baseline.

Trauma-Induced Stress Disorders, Including Response to Brain Injury

Resilience refers to the ability to successfully adapt to stressors, maintaining psychological well-being in the face of adversity. Neuroscience research has identified biomarkers for resilience, and studies have identified several attitudes and behaviors that foster psychological resilience to stress. Neuroscience has also identified risk factors associated with the development of PTSD. Evidence is increasing that stress disorders, including PTSD, are more common among soldiers than formerly believed.

The statement of task for the study specifically requested that the study not emphasize medical applications, so the committee focused on PTSD/TBI research that could be leveraged for nonmedical applications and that could lead to increased understanding of issues other than medical treatment per se. Nevertheless, the growing recognition that minimal to moderate brain traumas have chronic effects has long-term implications for future care requirements and associated costs. Neuroscience research into immediate care in combat areas, rehabilitation, new pharmaceutical treatments, and diagnostic tools can provide solutions to these problems.

Recommendation 10. The Army should support continued research on the identification of risk factors for the development of post-traumatic stress disorder (PTSD). This research could inform interventions that mitigate the risk for PTSD and related stress disorders, thereby lessening the performance deficits and disability resulting from these disorders.

Neuroscience research has identified risk factors associated with the development of PTSD and related stress disorders. The evidence is increasing that these stress disorders are more common among soldiers than was formerly believed.

Recommendation 11. The Army should apply the rapidly advancing understanding of the acute neuropathology of blast-induced traumatic brain injury, including the delayed neuropsychiatric effects of injuries as well. Mitigation strategies should include immediate postblast care using medication and/or other neuroprotective approaches proven to reduce the risk and severity of performance degradation. The Army should also continue its research in protective body armor.

Improving Cognitive and Behavioral Performance

Increased vigilance and enhanced perceptual discrimination, such as being able to recognize salient features or patterns, are inherently valuable to military missions. Research in a number of neuroscience subdisciplines, including computational neuroscience, systems neuroscience, and neuroergonomics, could lead to significant improvements in the cognitive skills of the soldiers and officers conducting Army operations.

Recommendation 12. The Army should structure its announcements of opportunities for research to draw broadly on multiple scientifically sound approaches to improving cognitive and behavioral performance, extending across the entire spectrum of neuroscience research rather than relying on a single approach. Army research opportunities should foster peer-reviewed competition and the synergism of collaboration across subdisciplines and approaches.

Neuroergonomics, which is an emerging field within the broader field of brain–machine interfaces, explores the ability of the brain to directly control systems beyond traditional human effector systems (hands and voice) by structuring the brain's output as a signal that can be transduced into a control input to an external system (a machine, electronic system, computer, semiautonomous air or ground vehicle, etc.). The Army Research Laboratory is now exploring the potential benefits of neuroergonomics. In the Army context, the goal of neuroergonomics is to facilitate a soldier–system symbiosis that measurably outperforms conventional human–system interfaces.

Recommendation 13. The Army should continue its focus on neuroergonomic research, using measured improvements in performance over selected conventional soldier–system interfaces as the metric to evaluate the potential of neurophysiology and other neuroscience disciplines in Army-relevant R&D for improving cognitive and behavioral performance.

TECHNOLOGY DEVELOPMENT OPPORTUNITIES

The committee identified and assessed cutting-edge and high-payoff technology opportunities, emphasizing their potential value for Army applications. Technologies were categorized as mission-enabling (directly enabling Army mission areas), research-enabling (supporting neuroscience-based research of high relevance to Army applications), or both.

To arrive at a set of high-priority investments, the committee assessed not only the potential value of prospective opportunities but also the time frame for developing an initial operational capability and the extent of external investment interest that the Army could leverage. Table S-1 lists the committee's recommended high-priority opportunities for Army investments in neuroscience.

Recommendation 14. The Army should invest in the high-priority technology opportunities listed in Table S-1. The investments should initially include long-term (5 or more years) commitments to each opportunity.

Table S-2 lists "priority" technology development opportunities that the committee recommends for limited Army investment. The committee viewed these opportunities as supplementing those in Table S-1 and recommended providing limited funding for R&D to explore potential applications.

Recommendation 15. The Army should consider limited investments (2 or 3 years for the initial commitment) in the technology opportunities listed in Table S-2. Evaluation of the results for each initial investment combined with assessment of outside progress in the field should guide decisions on whether to continue the funding for additional periods.

OVERARCHING RECOMMENDATIONS

The committee found two crosscutting issues that go beyond any particular request in the statement of task but that must be addressed by the Army if the potential benefits of neuroscience are to be fully realized.

A Mechanism for Monitoring New Opportunities in Neuroscience Research and Technology

The committee could find no single place in the Army science and technology structure from which progress in neuroscience, construed broadly, is being monitored for potential application by the Army and from which coordinating guidance can be disseminated to centers of neuroscience-relevant R&D around the country. This failure to identify and leverage advances in neuroscience is the most significant barrier to implementation of the 15 recommendations and is exacerbated by the diffusion of much of the R&D taking place in neuroscience.

Most of the opportunities listed in Table S-3, as well as those in Tables S-1 and S-2, involve areas of neuroscience that the Army needs to monitor for progress. Additionally, the committee also identified four important trends in neuroscience research:

- Discovering and validating biomarkers for neural states linked to soldiers' performance outcomes.
- Using individual variability to optimize unit performance.
- Recognizing opportunities from the vertical integration of neuroscience levels.

- Gaining new insights into the behaviors of adversaries.

Opportunities arising from the four research trends—and the many others yet to surface—will continue to revolutionize our understanding of the embodied mind and foster practical applications in civilian, commercial, and military affairs.

Neuroscience research and applications are advancing at a lightning pace, and the Army needs a reliable way to monitor progress in areas of nonmilitary neuroscience research and technology development. Direct Army investment in these areas will probably not be warranted unless an Army-unique application of substantial value emerges. Nonetheless, the Army should stay abreast of what is happening in these areas and have mechanisms in place to leverage the research results and adapt new technology for Army applications.

Recommendation 16. The Army should establish a group consisting of recognized leaders in neuroscience research in both the academic and private sectors to track progress in nonmilitary neuroscience R&D that could be relevant to Army applications. To ensure that the monitoring group remains sensitive to and abreast of Army needs, the membership should also include Army civilians and soldiers whose backgrounds and interests would suit them for meaningful participation in the group's deliberations.

Individual Variability as a Future Force Multiplier

A number of the recommendations reflect a common theme that may challenge traditional Army approaches but that offers great potential for increasing Army capabilities. Recommendations 2 (on training), 3 (on decision making), and 4 (on soldier stress response) all point to a larger theme that is emerging from current neuroscience research: Individual differences in behavior, cognition, and performance of skilled tasks are as deeply rooted in the neural structure of individuals as differences in strength, stamina, height, or perceptual acuity are rooted in their physiology. This common theme, as it pertains to opportunities for the Army to apply neuroscience, is explicitly explored in Chapter 8 of the report as a significant long-term research trend: using individual variability to optimize unit performance.

Neuroscience is establishing the role that neural structures play in the individual variability observed in cognition, memory, learning behaviors, resilience to stressors, and decision-making strategies and styles. Individual differences among soldiers have consequences for many Army applications and can influence operational readiness and the ability of Army units to perform assigned tasks optimally. Individual variability is in many ways at odds with the conventional approach of training soldiers to be interchangeable components of a unit.

Recommendation 17. Using insights from neuroscience on the sources and characteristics of individual variability, the Army should consider how to take advantage of the variability rather than ignoring it or attempting to eliminate it from a soldier's behavior patterns in performing assigned tasks. The goal should be to seek ways to use individual variability to improve unit readiness and performance.

1

Introduction

The Army has made great strides in exploiting technological advances on the battlefield. This success is based in large part on advances in computers and miniaturization, which exploit reductions in scale and exponential increases in performance, and on advances in disciplines such as information science and network science. The field of neuroscience offers similar potential to achieve further improvements in soldier performance for future operations.

Advances and major investments by the broader community in neuroscience promise new insights for military applications. These include traditional areas of importance to the Army, such as learning, decision making, and performance under stress, as well as newer areas, such as cognitive fitness, brain–computer interfaces (an extension from earlier human–computer ergonomics), and biological markers of neural states. Advances in such fields as functional magnetic resonance imaging (fMRI) and bioengineering have resulted in instrumentation and techniques that can better assess the neural basis of cognition and enable visualization of brain processes. These have the potential to provide new measures of training and learning for soldiers, while also shedding new light on traditional approaches to behavioral science used by the Army. Continuing research is certain to give rise to new opportunities, and the Army would like to better understand how these neuroscience opportunities can be exploited for the benefit of the soldier.

STUDY BACKGROUND

Most current neuroscience research is at a basic level with little or no regard for longer-term military potential. In recent years, however, it has begun to capitalize on the investments in basic research and move toward applications. The time is right to apply neuroscience understanding to applications that have military relevance. What is needed is a determination of the specific outcomes of this basic research that are likely to lead to the development of neurotechnologies with possible military application and a discussion of the spectrum of efforts under way, with emphasis on the nonmedical applications.

In March 2007, the National Research Council's (NRC's) Board on Army Science and Technology (BAST) convened a meeting that attempted to bring the complexities of neuroscience and its military applications into focus. Presentations to the BAST on ongoing work in academia, industry, and government included areas of science and technology at the intersection of diverse fields and speculated on possible synergies between ongoing research efforts.

The Army believes that neuroscience will grow to impact numerous applications that are presently scattered among multiple disciplines and fields. It is interested in identifying a range of potential applications and bringing coherence to an emerging collection of relevant neuroscience advances that can serve as a basis for future investments in research.

Statement of Task

The Assistant Secretary of the Army (Acquisition, Logistics, and Technology) (ASAALT) requested that BAST conduct a study of the potential of neuroscience to support military applications. Box 1-1 contains the statement of task for the study.

The Army sponsor requested that the study address what neuroscience can be expected to do as well as what advances could be made if appropriate direction is provided and investments are made. Because the field of neuroscience is wide-ranging and other entities are investing large amounts in many neuroscience subdisciplines, the study should focus on opportunities that could have a high-payoff potential for the Army and on areas where it is unlikely that others will devote substantial resources that will benefit Army applications. The study should also suggest opportunities for leveraging specific investments by others, where appropriate, for Army applications. The sponsor specifically requested that the study consider opportunities achievable over the next 5, 10, and 20 years and avoid unrealizable "bionic soldier" applications.

BOX 1-1
Statement of Task

The Assistant Secretary of the Army (Acquisition, Logistics, and Technology) (ASAALT) has requested the NRC BAST to conduct a study of neuroscience in terms of its potential to support military applications. The study will address what neuroscience can be expected to do as well as what neuroscience advances could do if provided appropriate direction and investment. Given the fact that the field of neuroscience is very extensive and there are many other investments underway in numerous areas and sub-areas, this study will focus on those areas that have high-payoff potential for the Army where it is unlikely that others will devote substantial resources to research and exploitation in these areas for Army benefit. The study will also suggest opportunities for leveraging specific investments where appropriate for the Army. Specifically, the study will

1. Identify and recommend novel technologies, methodologies and approaches for assessing and guiding the training of Army personnel to enhance soldier learning. The study will consider:

- Assessing how neural pathways implicated in functional processing can be enhanced to improve the training of Soldiers in an operational context
- Examining how sleep deprivation and high stress conditions influence training efficiency and effectiveness through degradation of specific neural pathways involved in learning and memory
- Describing how neural pathway approaches can be applied by the Army to more objectively assess training paradigms, including virtual reality training as compared with constructed reality and operational conditions, regarding their efficacy in improving performance by Soldiers in combat environments
- Whether traditional behavioral science as applied to Soldier training, learning and performance can benefit from developments and new knowledge being acquired in areas of neuroscience that have significant potential to impact the Soldier.

2. Examine leading-edge methodologies and technologies developed in the government, private and academic sectors to improve cognitive and behavioral performance, particularly under high stress conditions. Consider representative non-military task environments requiring continuous operation with high vigilance and risk.
3. Identify additional high-risk, high-payoff opportunities in the neuroscience field with strong potential for Army application. Identify critical barriers (such as legal and ethical) to research and development that could be surmounted by appropriate science and technology investments assuming that these are Army critical and unique. Suggest ways to overcome the barriers, and recommend research initiatives. Identify areas and opportunities where the Army can leverage relevant investments of others for Army application.
4. Determine trends in research and commercial development of neuroscience technologies that are likely to be of importance to the Army in the longer term.

Study Approach

The NRC appointed the Committee on Opportunities in Neuroscience for Future Army Applications to carry out the study. Special care was devoted to the composition of the committee. Some members had backgrounds in the traditional facets of neuroscience such as psychology and cognitive science and in neurology, including neuronal stimulation, neuropharmacology, imaging techniques, and human–computer interfaces (traditional ergonomics); others had expertise in newer and emerging subdisciplines and cross-disciplinary fields such as neuroimmunology, neuroeconomics, neuroergonomics, augmented reality, and computational neuroscience. Members were also selected on the basis of their experience in research and development (R&D), military operations, and medicine, and in training specialties such as memory and learning, assessment, decision making, prediction, and reading intentionality. Short biographies for the members are given in Appendix A.

Initially the committee was divided into data-gathering teams based on the Army's own perception of neuroscience requirements. The teams determined sources of outside expertise that would be helpful to the committee's study and reviewed recent publications on neuroscience topics, including two recent NRC studies for DOD sponsors (NRC, 2008a, 2008b). The streams of data-gathering activity were brought together midway though the committee's deliberations when the first full-message draft was being written. A consensus was reached on pertinent findings to be contained in the report, and the committee was reconstituted into writing teams to draft the findings. The committee's conclusions and recommendations were refined and ratified at the final meeting. All of the meetings and data-gathering activities are documented in Appendix B.

REPORT ORGANIZATION

This report contains the committee's analysis, conclusions, and recommendations. It focuses on areas of neuroscience research likely to lead to developments of interest to the Army and provides specific objectives for the Army to consider.

Chapter Structure

Chapter 1, the Introduction, provides the study background and report organization. Chapter 2, Neuroscience and the Army, provides a brief history and definition of neuroscience, discusses Army applications likely to be served by neuroscience advances, and covers issues related to such advances. Chapter 3, Training and Learning, discusses the assessment and testing of soldiers and units, and Chapter 4, Optimizing Decision Making, considers the multiple roles of leaders and tools for characterizing and predicting behaviors. Chapter 5, Sustaining Soldier Performance, discusses degradation of performance as a consequence of exposure to various environmental stressors (e.g., fatigue, metabolic stressors, pain, and sleeplessness), countermeasures to these stressors that aim to prevent such degradation or restore baseline performance, including pharmacological approaches, and countermeasures to the longer-term neural and cognitive effects of brain injury and traumatic stress. Chapter 6, Improving Cognitive and Behavioral Performance, describes emerging approaches to enhancing soldier performance that combine neuroscience insights with cognitive-behavioral ergonomics; one such approach would come from the new field of neuroergonomics. It also assesses the potential utility to the Army of pharmacological and behavioral interventions to enhance cognitive performance. Chapter 7, Neuroscience Technology Opportunities, assesses high-risk, high-payoff technology opportunities in terms of their potential importance to the Army, the likelihood of their development by others (leveraging opportunities), and the time frame for initial operational capability. Chapter 8, Long-Term Trends in Research, describes major trends in neuroscience research that are likely to yield future opportunities for the Army and should therefore be monitored by a suitable and continuing mechanism. Finally, Chapter 9, Conclusions and Recommendations, presents the committee's specific conclusions and recommendations in response to the statement of task, plus overarching conclusions and recommendations that follow logically from the specific recommendations.

Response to Statement of Task

The chapter structure of the report does not correlate in a simple way with the four subtasks in the committee's statement of task (see Box 1-1). This section describes which parts of the report address each subtask.

Subtask 1 requests that the committee identify technologies, methodologies, and approaches applicable to training Army personnel, including (1) assessing training paradigms and improvements to training; (2) examining the influence of high stress and sleep deprivation; and (3) determining how traditional behavioral science approaches benefit from new knowledge in neuroscience. Chapter 3 responds to the first item by discussing training paradigms and methods, performance assessments of individuals and groups, identification of training candidates, and metrics for training effectiveness. The chapter also addresses the third item by describing the impact of neuroscience advances on traditional behavioral approaches to the assessment of both training and performance. The second item—the influence on soldier performance of high stressors, such as fatigue, pain, and sleep deprivation—is discussed in Chapter 5.

Subtask 2, which requests an examination of methodologies and technologies to improve cognitive and behavioral performance, is addressed in Chapters 4, 6, and 7. Specifically, Chapter 4 discusses methodologies for studying decision making and tools to predict how an individual will approach making a decision. It suggests ways to capitalize on individual variability and deal with the constraints of a decision maker's belief system on the decisions he or she makes. As discussed in Chapter 6, improvements in cognitive and behavioral performance are also likely to arise from the emerging field of neuroergonomics, as well as from research by the auto industry and NASA, which are doing research that is also of interest to the Army. The committee recognizes the potential for neuropharmacological approaches to improving performance, but in Chapter 6 urges caution because there may be unknown side effects and long-term consequences of using pharmacological agents for this purpose. Technologies that might be used to improve mission-related performance are assessed in Chapter 7. The opportunities include field-deployable indicators of neural state and advances in human–machine interfaces and brain–computer interfaces such as augmented reality, three-dimensional haptic interfaces, and information management to cope with cognitive overload.

Subtask 3 asks the committee to identify high-risk, high-payoff opportunities in neuroscience, critical barriers to R&D, and areas where the Army can leverage the investments of others. High-payoff research opportunities are identified in the recommendations from Chapters 3 through 6; these opportunities vary in the level of risk associated with them, as discussed in the respective chapters. Chapter 7 describes technology development opportunities, and Tables 7-1 and 7-2 summarize the committee's evaluations of high-priority and priority opportunities for Army investment, taking into account risk-benefit trade-offs and the just-mentioned potential to leverage investments by others. Legal and ethical barriers to implementing neuroscience research results (and technology) are discussed in Chapter 2, while technical barriers to implementing technology opportunities are discussed in Chapter 7.

Subtask 4 asks the committee to identify trends in research and in the commercial development of neuroscience technologies that are likely to be important to the Army in the longer term. Chapter 7 includes a section on long-term trends in technology. Chapter 8 discusses long-term neuroscience research trends that the Army should monitor for results important to Army missions. In Chapter 8, the committee suggests a mechanism for effective monitoring of the many disciplines and areas of research and technology development in neuroscience that are likely to produce results of value to the Army over the longer term.

REFERENCES

NRC (National Research Council). 2008a. Emerging Cognitive Neuroscience and Related Technologies. Washington, D.C.: The National Academies Press.

NRC. 2008b. Human Behavior in Military Contexts. Washington, D.C.: The National Academies Press.

2

Neuroscience and the Army

This chapter provides a brief overview of neuroscience and neuroscience technology, including the definition of neuroscience used by the committee for the study. It describes major applications areas that were considered by the study and provides rationale for not considering some applications.

HISTORY, SCOPE, AND DEFINITION OF NEUROSCIENCE

"Neuroscience" refers to the multiple disciplines that carry out scientific research on the nervous system to understand the biological basis for behavior. Modern studies of the nervous system have been ongoing since the middle of the nineteenth century. Neuroanatomists studied the brain's shape, its cellular structure, and its circuitry; neurochemists studied the brain's chemical composition, including its lipids and proteins; neurophysiologists studied the brain's bioelectric properties; and psychologists and neuropsychologists investigated the organization and neural substrates of behavior and cognition.

The term "neuroscience" was introduced in the mid-1960s, signaling the start of an era when these disciplines would work together cooperatively, sharing a common language, common concepts, and a common goal: to understand the structure and function of the normal and abnormal brain. Neuroscience today spans a wide range of research endeavors, from the molecular biology of nerve cells, which contain the genes that command production of the proteins needed for nervous system function, to the biological bases of normal and disordered behavior, emotion, and cognition, including the mental properties of individuals as they interact with each other and with their environments.

Neuroscience is one of the fastest growing areas of science, and the brain is sometimes referred to as the last frontier of biology. In 1971, for example, the first meeting of the Society for Neuroscience was attended by only 1,100 scientists; in 2007, 26,000 scientists participated at the society's 37th annual meeting and more than 15,000 research presentations were made. At this time, national societies of neuroscience exist throughout the world, and there exists a Federation of European Neuroscience Societies.

Neuroscience incorporates a number of interacting areas, including cognitive neuroscience, systems neuroscience, cellular and molecular neuroscience, developmental neuroscience, clinical neuroscience, theoretical neuroscience, and computational neuroscience. Operations involving neurons[1] form the basis for all of these areas and take place on four fundamental hierarchical levels: molecular, cellular, systems, and behavioral. These levels rest on the principle that neurons communicate chemically by the activity-dependent secretion of "neurotransmitters" at specialized points of contact called "synapses." In order for the brain to perform its multiple functions, the mental operations of the brain rely on a properly functioning and integrated system of autonomous bodily functions, monitored by the brain and modified by behavioral operations only when the autonomous regulatory systems—thermal control, control of blood nutrients, orientation of the body and limbs in space while moving, gesturing, and transporting, control of the salt and water balance, and so on—are compromised.

The autonomous peripheral systems monitored by the brain and over which it has ultimate control allow for an extreme but generally subconscious interaction between the body's physical fitness and the brain's emotional and cognitive powers to drive the body's performance under demanding conditions. For example, overcoming fatigue through personal willpower alone is a learnable ability that is characteristic of superior athletes and that would be beneficial for military personnel as well (see Chapter 5).

Many fields of clinical medicine are directly concerned with the diseases of the brain. The branches of medicine most closely associated with neuroscience from the perspectives of this report are neurology (the study of the degenerative

[1]Nerve cells of the central and peripheral nervous systems.

sensory and motor diseases of the brain), neurosurgery (the study of the surgical treatment of neurological disease), and psychiatry (the study of behavioral, emotional, and mental diseases). Other fields of medicine also make important contributions to neuroscience, including neuroradiology, which is the use of radiation for imaging the brain—initially with X-rays and, more recently, with positron emitters, radiofrequency, and electromagnetic waves—for clinical studies and microscopic study of samples from diseased neural tissue.

Hierarchical Levels of Neuroscience

At the molecular level, one examines the interaction of molecules—typically proteins—that regulate gene expression and translation into proteins. Proteins mediate neurotransmitter synthesis and storage and release other essential neuronal molecular functions such as the receptors by which neurons respond to neurotransmitters. Most drugs used for the treatment of neurological or psychiatric diseases work by either enhancing or diminishing the effects of neurotransmitters.

At the cellular level of neuroscience, one examines the interactions between neurons through their synaptic connections and between neurons and the supporting cells, the glia. Research at the cellular level strives to determine the neural pathways by which specific neurons are connected and which of their most proximate synaptic connections might mediate a behavior or behavioral effects of a given experimental perturbation.

At the systems level, one examines the interconnected neural pathways that integrate the body's response to environmental challenges. The sensory systems include the specialized senses for hearing, seeing, feeling, tasting, and balancing the body. The motor systems control trunk, limb, eye, and fine finger motions. Internal regulatory systems are responsible for, among other things, control of body temperature, cardiovascular function, appetite, and salt and water balance.

At the behavioral level of neuroscience research, one examines the interactions between individuals and their collective environment. Research at this level centers on the systems that integrate physiological expressions of learned, reflexive, or spontaneous behavioral responses. Behavioral research also looks at the cognitive operations of higher mental activity, such as memory, learning, speech, abstract reasoning, and consciousness. Research over the past three decades has established that the brain is highly adaptable (this ability is commonly termed "neuroplasticity") at each level of operation: the activity-dependent ability to change gene expression, to change transmitter production and response, to change cellular structure and strength of connections between neurons, and to change behaviors by learning.

An important consequence of organizing neuroscience research at four vertical hierarchical levels is that it enables one to hypothesize experimental results on one level based on experimental findings and observations from other levels. This ability extends to hypothesizing neuronal operations or neuronal diseases based on data that would predict results at the behavioral level given the results of a perturbation or other experimental manipulation at a lower level. Such results are strongly supported in the literature (Aston-Jones and Cohen, 2005a, 2005b).

One might predict from experimental results in animals, for example, that the thin axons that establish functional properties of the noradrenergic system might also be one of the brain fiber systems most vulnerable to the percussive damage of traumatic brain injuries (TBI), such as might result from an improvised explosive device (IED). As discussed in the Chapter 5 section on brain injury, this is indeed the case, and the ability to translate between levels of neuroscience has proven helpful in the treatment of TBI and its emotional effects.

NEUROSCIENCE TECHNOLOGIES

Until the advent of modern computer-based technology, the primary noninvasive tools used to understand the workings of the central and peripheral nervous system were the recording of electrical signals from the scalp (electroencephalography [EEG]) and X-ray imaging of the soft tissue of brain as distinguished from bone and compartments containing cerebral spinal fluid (CSF). EEG allowed detecting epileptogenic foci that could subsequently be managed surgically if a discrete region was involved in the initiation of seizures or pharmacologically if the region was more generalized. The X-ray imaging allowed detection and localization of lesions because the lesions displaced readily identified portions of the brain. However, these technologies provided very limited insight into neural information processing related to cognition, the central mechanisms involved in the perception of pain, or other higher-order brain activities. The pioneering work of Penfield and his colleagues was an exception: It combined EEG with invasive brain surgery to associate the visual and auditory auras that accompanied seizures to specific regions in the visual, auditory, or temporal cortices (Penfield and Perot, 1963).

The two decades from the late 1980s to the present have seen the rapid rise of technologies that can provide high-resolution structural images of the gray and white matters of brain as distinct from one another, clearly delineating details as small as the foci of white-matter disease and inflammatory changes. These technologies are capable of imaging the metabolic processes that are associated with functional activity of the brain in response to specific stimuli (positron emission tomography [PET] and functional magnetic resonance imaging [fMRI]); the orientation and dimensions of axonal fiber bundles connecting one brain region to another (diffusion tensor imaging); and the electrophysiological localization of brain activation (magnetoencephalography

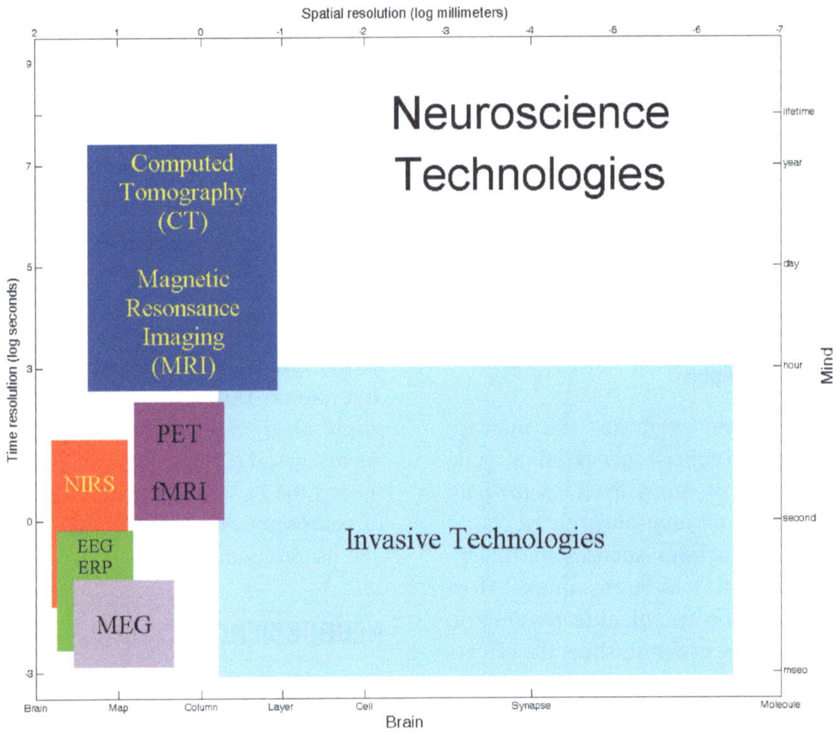

FIGURE 2-1 Various noninvasive imaging technologies provide insight into the brain (anatomy) and mind (function). Resolution boundaries are approximate. Currently, high resolution requires invasive procedures or injection of pharmacological agents not approved by the Food and Drug Administration; such technologies are outside the scope of Army applications, but some discussion is provided in Chapter 7. Note that useful measurements are performed at any point in the brain-mind plane. SOURCE: Adapted from Genik et al., 2005.

[MEG] and visual and auditory evoked potentials). Most of the newer methodologies depend on computer-driven signal averaging across a time span of seconds to provide an understanding of brain structure and function.

The advancement of noninvasive neural measurement techniques has opened new windows to the study of the functioning human brain. Figure 2-1 illustrates how neuroscience technologies provide insight into the brain (anatomy) and the mind (function). The spatial resolution of a given technology defines the largest and smallest brain structures that can be observed, while the temporal resolution defines the elements of mind function to be measured. Academic and commercial research are primarily geared to improving resolution, although important measurements for the prediction of behavior can be made at any point in the brain-mind plane. Chapter 7 contains further details of neuroscience technologies, including a discussion of more invasive modalities and the future direction of noninvasive imaging research.

The spatial resolution of the structural imaging modalities that could be achieved with the human brain in 2008 approximates 100 microns per side in three-dimensional pixel elements when a 7-tesla (T) MRI system is employed, which is the highest-field whole-body scanner currently in commercial production. This resolution requires a total imaging time of less than 10 minutes in a series of planes. Recognizing that a large neuron cell body has a diameter of 20-30 microns, each 100-micron element is likely to contain eight neurons and a much larger number of glia. Also recognizing that each dendrite of a large neuron (Purkinje cell or motor neuron or pyramidal cell) may contain 10^3 to 10^4 synaptic connections, the number of information-processing elements contained within the small pixel volume is large. There are very few 7-T MRI instruments available for human imaging at this time. The majority of MRI machines are 1.5 T and 3 T. The resolution for these instruments nominally decreases proportionally with static field, assuming a similar image acquisition time. Longer imaging times can yield higher resolution, but the trade-off is that any movement of the head by the subject markedly decreases the resolution. The limits of structural resolution are therefore spatiotemporal limitations.

The diffusion tensor imaging allows examining the pathways along which groups of axonal fibers travel from one brain region to another and the vectorial direction of the fiber bundle group. This information is essentially a mapping of the potential for information exchange between regions and the tracing of the bundles that could carry information. No information is provided about the actual informational transactions between brain regions or the temporal events associated with such transactions.

The MRI system depends on alignment of the bulk water dipole in the magnetic field and the perturbation of this alignment by a radiofrequency pulse. The realignment of the dipole following the radiofrequency pulse provides the signal of interest. What MRI then permits resolving is the relative water content of each compartment in the brain, spinal cord, or organ system that is being studied. The white matter has the least free water content, the gray matter has intermediate levels, and the CSF compartments have the greatest amount of water. These compartments can be readily resolved. The loss of myelin accompanying multiple sclerosis or other white-matter disease is also readily discerned for this reason, as are other diseases due to neurodegenerative changes and cell loss. (MRI is to soft tissue as a computerized axial tomography [CAT] scan is to bone. These methods are essentially noninvasive.)

Imaging of the brain based on X-ray methods (CAT scans or X-ray scans) also have a high resolution but do not readily differentiate white from gray matter as does an MRI device. This approach has proven to be very useful for threading catheters intravascularly to within 1 mm of any location in the brain, for localizing tumors in the cranial cavity and region of the spinal cord, and for visualizing new blood from hemorrhages. The two modalities taken together offer exquisite timing of tumor, infection, and bleeding developments.

The technologies for structural imaging provide the basis for understanding the brain regions of interest obtained by functional images. Functional images are lower resolution than structural images in order to facilitate acquiring many repetitions in a short time. Typical 3-T fMRI acquires 2- to 3-mm isotropic resolution whole-brain images every 1 or 2 seconds, while cutting-edge hardware can acquire 1-mm isotropic resolution whole-brain images in 1 second. By acquiring the images in a time-locked fashion with stimulation, one can interpret the activation of brain regions as a response to stimulation. The fMRI signal is generated by the loss of oxygen from hemoglobin (the deoxygenated hemoglobin is paramagnetic, while the oxygenated is not) and the change in blood flow in the activated region. The blood flow response continues for several seconds, enabling detection of very fast physiological events using this rather slow temporal sampling. This is called blood oxygen level-dependent (BOLD) imaging.

The more repeatable the brain response and, thus, the blood flow response, the better localized activations can be observed, because one can average over several trials. If one can constantly stimulate a response in a block of time, say the visual cortex with a flashing checkerboard, an activated state lasts longer and is easier to detect. However, some cognitive processes are inherently transient and are better observed in so-called event-related experiments. Which experimental method to use depends on the physiology of the reaction one wishes to study (Friston et al., 1999; Otten et al., 2002).

In PET imaging, a solution containing a small amount of a radioactive element is injected into the subject's bloodstream. Signals from the decaying isotopes will show localized concentration, which could be due to increased blood flow, say, or perhaps to glucose metabolism, depending on the tracer used. PET functional images are acquired at 3- to 4-mm isotropic resolution, and because very little tracer is used (to protect the subject), they require a few minutes to tens of minutes of averaging to show a significant signal, again, depending on the tracer used. The significant advantage of PET over fMRI is that specific metabolic reactions can be targeted for measurement rather than merely observing changes in blood flow. Additionally, with proper calibration, an absolute measure of change in brain function can be determined, even in different scanning sessions, whereas with fMRI only changes relative to a baseline condition are observable. The main disadvantage of PET is that the physiology that produces transient signals of short duration, akin to noise, cannot be observed. Moreover, to avoid any adverse effects of radiation on the subject, PET studies on a given individual are limited to three or four experiments per year depending on the amount of emitter injected. The functional properties of the brain have been shown to have relevance for detecting a loss of situational awareness in individuals subjected to total sleep deprivation for more than 24 hours (Belenky et al., 2003).

Some of the noninvasive imaging technologies have characteristics that may limit their use in military applications. PET, for example, requires injection of a positron emitter into the human subject. On the other hand, fMRI is truly noninvasive in that no contrast agents or radioisotopes are injected in the subject.

A recent study by Bakker et al. (2008) demonstrated that activation of the fMRI in hippocampal regions was related to the detection by human subjects of minor changes to objects in the visual field. Detection of changes in the visual field by an unconscious mind might be applied to the identification of IEDs implanted in the terrain. A clear understanding of the mechanisms by which an individual can become alerted to such activations when hippocampal neurons fire might enhance the mission success of Army personnel.

Electrophysiological events accompanying brain stimulation can be recorded in real time (milliseconds) from the cortex but not from deep brain structures, and the spatial resolution is 5 mm per recording element. For noninvasive studies with the calvarium intact, the recordings are essentially averaging events in the top millimeter of cortex and across a diameter of 5 mm or more. The evoked potential is a signal-averaging methodology that examines rates of event-related depolarization across spatial domains of the scalp associated with a specific stimulus presentation—for example, a checkerboard flicker, a flash, or an auditory signal. It is possible to detect changes in such events that are due to changes in information processing (or to central nervous system lesions). Because vectorial electrical events are

BOX 2-1
Computational Processes in the Human Brain

The last half century has seen the emergence of multiple technologies that, when taken in total, can lead to new understanding of the human–machine interface and to improving our management of it as an interacting organism. The relatively concurrent development of high-speed computer data processing, anatomic mapping of the human brain, noninvasive imaging technology (capable of visualizing neural information processing at a resolution of 1 mm^3 in three dimensions), decoding of the human genome, and nanotechnology (giving us the ability to prepare molecular entities that can self-assemble in predictable arrays) has permitted researchers and theoreticians to construct models of how human–machine interfaces can aid training, increase combat effectiveness, and speed up the acquisition of information by the mind. A summary of some of this can be found in Kurzweil (2005, pp. 122-128).

Anatomic studies of the brain lead to an estimate that the brain contains about 10^{11} neurons (information-processing units) and that many neurons (Purkinje cells in the cerebellum or anterior motor neurons in the spinal cord) have 1,000 information input elements (synapses) and 10 output elements (synapses), yielding about 10^{14} possible information transactions occurring in seconds (it is assumed that in a local circuit in the cerebellum there are rapid impulse and repolarization events). If we focus on the neural components that enable all human visual perception, the retina is the initial image capture and processing element. It is a structure about 2 cm wide by 0.5 mm thick and weighs about 20 mg. This volume of tissue will contain about 10^5 neurons. Local information processing occurs between the rod and cone cells, which are activated by the input from the visual field, and the ganglion cells, which transmit initially processed information to the occipital cortex of the brain for further processing.

The retina provides initial processing capability such as rapid detection of edges of objects by center-surround inhibition/activation, the movement of elements in the visual field, and the perception of dark/light properties of the objects in the field. The range of light/dark information processing that can be detected is several orders of magnitude. During the next two decades it is likely that visual input devices will be developed that will be able to fuse visual data with auditory cues to allow the more rapid detection of threats. These devices will be able to sustain vigilance by fusing measures of decreases in situational awareness (obtained through electroencephalography (EEG), evoked potentials, or other cues from the soldier) with threats detected in the area of operations.

accompanied by changes in the magnetic field perpendicular to the electrical path, it is possible to measure columnar electrical events in the sulci using MEG.[2] The spatial resolution with this method is on the order of centimeters, and the temporal resolution is on the order of fractions of seconds to seconds. Studies on the effects of various stressors on cognitive ability using MEG with EEG provide evidence that this tool can detect decremented situational awareness. This is important from a defense perspective since EEG-related devices are portable and readily deployable on an individual soldier while MRI devices are not (Box 2-1).

Among the applications of modern neuronal imaging and structural technologies to military needs are the assessment of (1) when appropriate training-to-criterion on a given skill set has been achieved; (2) when a soldier in the well-rested state has significantly degraded situational awareness of her/his capability; (3) early signatures of neural dysfunction. While some of these methodologies may prove to have application for the a priori assessment of highly desirable traits such as leadership, persistence, and other successful warrior behaviors, there is little evidence to support application of the neuroscience tools currently available to tasks such as the three mentioned above.

The report thinks of technology as being in one (or sometimes both) of two categories: technologies that are "mission-enabling" (deployable) instruments and those that are "research-enabling" instruments. The word "instrument" is used in the most general sense: It could refer to a pen-and-paper personality inventory, a software-controlled skills survey, a reaction-time analysis method for training assessment, a control interlock system to distribute information to vehicle crew based on their current workload and baseline cognitive capability, an in-helmet device designed to monitor neural activity or cerebral blood flow, or a device that advances an imaging technology. Neuroscience research plays a major role in the development of instruments in both categories of technology.

RELIABLE BIOMARKERS FOR NEUROPSYCHOLOGICAL STATES AND BEHAVIORAL OUTCOMES

Recordings of the brain's electrical activity and the images of neural activation regions that were described in the previous section on neural monitoring technologies are all methods to record changes in brain activity in a specific location at a given time. Generally, these data on neural events are interpreted in terms of what they tell us not just

[2]This is possible because in the cerebral cortex, information-processing modules are composed of a column of neurons spanning the thickness of the cortex.

about that event but also about longer-lasting patterns of brain activity and brain structure—that is, about neural states and changes in neural state. A major, if not the primary, goal of studying neurophysiology with these techniques is to understand the linkages among neural states, psychological states, and behavior. (Psychological states are traditionally self-reported; behavior is directly observable by others.) In short, the outputs of neural monitoring technologies are indicators of neural state (and of changes in neural state) that can in turn be linked with behavior.

Typically, these linkages begin with relationships that have statistical significance on a group-averaged basis and move, as the state of scientific knowledge progresses, to connections that hold on an individual basis.[3] Ultimately, this refinement will lead to a reliable, scientifically defensible, knowledge of necessary and sufficient causal conditions underlying and explaining the observed patterns of brain activity, mental experience, and behavior. But because that ultimate goal is still a long way off, care must be taken not to leap prematurely from a statistically significant correlation to conclusions about causality.

In addition to brain activity signals and images, the neurosciences and allied fields—ranging from genetics to molecular biology and traditional behavioral science—are exploring a wide range of phenomena that can be connected with neuropsychological states or changes in state. Just as a neuroimaging pattern can be used as an indicator or marker of neuropsychological state, so may other phenomena. Among the phenomena being studied for this purpose are biologically active small molecules, proteins and related molecules (e.g., lipoproteins and metabolic residues or precursors of proteins), genes and nonencoding regions of the genome, physiological events or patterns outside the brain but within the organism, and responses to an environmental exposure (physical, chemical, biological, social, or psychological). The variety and complexity of the hypotheses being put forward and tested about such correlations are driving a revolution in scientific understanding.

"Biomarker" is a term often used in the biomedical disciplines for a characteristic that can be used as an indicator of some biological condition or outcome that is ultimately of interest but difficult to ascertain directly, at least under conditions of interest to a particular application. A number of implicit and explicit definitions of "biomarker" are in common circulation.[4] There are also quite different uses of the term in other disciplines.[5] To avoid confusion, the committee has adopted the following definition, published by the Biomarkers Definitions Working Group of the National Institutes of Health (Atkinson et al., 2001, p. 91):

> *Biological marker (biomarker)*: A characteristic that is objectively measured and evaluated as an indicator of normal biological processes, pathogenic processes, or pharmacologic responses to a therapeutic intervention.

In its report, the Biomarkers Definitions Working Group focused on applications of biomarkers as surrogates for clinical end points in a study or clinical trial. This emphasis on a biomarker's role as a surrogate for a physiological or behavioral condition or outcome is evident in the Working Group's definition of clinical end point as "a characteristic or variable that reflects how a patient feels, functions, or survives." Common examples of biomarkers mentioned by the Working Group include elevated blood glucose concentration for the diagnosis of diabetes mellitus, the concentration of prostate-specific antigen in blood as an indicator of the extent of prostate tumor growth and metastasis, and blood cholesterol concentration as a predictive and monitoring biomarker for heart disease risk (Atkinson et al., 2001, p. 91).

Throughout the remainder of this report, the Committee on Opportunities in Neuroscience for Future Army Applications is primarily interested in biomarkers as objectively measured and evaluated indicators of either a neural state or a behavioral outcome. For example, Chapters 3 and 5 discuss the use of neuroimaging as a source of biomarkers for individual response to particular environmental stressors. Chapter 7 discusses the value for Army applications of finding biomarkers that can be measured under field conditions and that are reliable indicators of specific neural states that have been reliably linked to behavioral outcomes. These biomarker applications differ from uses of biomarkers as surrogate clinical end points, which were the focus of the Biomarkers Definitions Working Group. Nevertheless, the Working Group's caveats about biomarker applications are useful cautions for any application.

For example, both the accuracy and precision of a biomarker as a surrogate measure of outcome must be demonstrated:

[3]For example, fMRI investigations typically begin by examining relationships averaged over many trials per subject and then averaged over multiple subjects. Once a statistically significant relationship is established in this way, the typical next step is to show that the relationship holds for individual events in the group of subjects, and ultimately to individual events in each subject.

[4]The National Cancer Institute defines a biomarker as "a biological molecule found in blood, other body fluids, or tissues that is a sign of a normal or abnormal process, or of a condition or disease . . ." in its *Dictionary of Cancer Terms*, available at http://www.cancer.gov/dictionary/. Accessed November 23, 2008. The MedlinePlus online medical dictionary uses a definition from Merriam Webster, in which a biomarker is "a distinctive biological or biologically derived indicator (as a biochemical metabolite in the body) of a process, event, or condition (as aging, disease, or exposure to a toxic substance)." The usage example given is "age-related biomarkers of disease and degenerative change." The URL for this National Institutes of Health (NIH)-sponsored dictionary is www.nlm.nih.gov/medlineplus/mplusdictionary.html. Accessed on November 23, 2008.

[5]In petroleum exploration, biomarkers are compounds found in geologic extracts (including oil, rock, sediment, and soil extracts) that indicate a biological origin of some or all of the material. See "Using Oil Biomarkers in Petroleum Exploration," by Oiltracers LLC, available at www.oiltracers.com/biomarker.html8. Accessed November 21, 2008. A similar use occurs in planetary science and astrobiology, where a biomarker is a chemical that signals the presence of biological processes, often in the distant past.

> The utility of a biomarker as a surrogate endpoint [or as a surrogate for a neural state or behavioral outcome] requires demonstration of its accuracy (the correlation of the measure with the clinical endpoint [or the neural state or behavioral outcome]) and precision (the reproducibility of the measure). (Atkinson et al., 2001, p. 92; bracketed text added by the committee)

Elsewhere in its report, the Biomarkers Definitions Working Group notes as follows:

> Biomarkers that represent highly *sensitive and specific* indicators of disease pathways have been used as substitutes for outcomes in clinical trials when evidence indicates that they predict clinical risk or benefit. (Atkinson et al., 2001, p. 90; emphasis added by the committee)

The attributes of sensitivity and specificity have rigorous definitions that can be applied across the range of physical, biological, and even social characteristics that are or will be candidates for indicators of whether a condition such as a neural state is present or not. In a binary test (in this case, whether the neural state of interest is or is not present), "sensitivity" is defined mathematically as the number of test instances in which the biomarker is positive and the neural state is present, divided by the number of instances in which the neural state was present whether or not the biomarker was positive. In other words, the measure of sensitivity is the ratio of true positive tests to the sum of the true positive and false negative tests. (False negative tests are those that should have been positive.) "Specificity" is defined as the number of instances in which the biomarker was negative and the neural state was absent, divided by the number of instances in which the neural state was absent whether or not the biomarker was present. The measure of sensitivity is thus the ratio of true negative tests to the sum of the true negative and the false positive tests. A reliable biomarker for this kind of binary application is one that has both high sensitivity and high specificity; that is, both ratios are close to unity.

With respect to how biomarkers may be used, they can be current, retrospective, or predictive (prospective) indicators or measures, depending on whether the condition or end point for which they are used as a surrogate occurs at the same time, before, or after the assessment of the biomarker. Demonstrating the reliability of a biomarker for an application typically requires the same temporal relationship as the intended application.

As the report by the Biomarkers Definitions Working Group also notes, often several biomarkers must be combined to get a reliable indicator or measure of outcome (Atkinson et al., 2001, p. 93). For applications of practical value to the Army, such as field-deployable indicators of neural state, this approach, assessing multiple biomarkers, may often be required (see Chapter 7 for further discussion).

In summary, applications of biomarkers as surrogates for neural states or behavioral outcomes require a demonstration of reliability that exceeds mere statistical correlation. In cases where the biomarker is a measured quantity that correlates with a magnitude of outcome, that quantitative relationship must be accurate and reproducible under the conditions in which the biomarker will be assessed in practice. In cases where the application involves a binary test (the outcome or end point to be indicated either is or is not present), values of sensitivity and specificity close to unity are required for reliability.

ARMY APPLICATION AREAS

Neuroscience represents at once both a challenge and a great opportunity. It is a challenge because the breadth and complexity of contemporary neuroscience are so great; and it is an opportunity because neuroscience can arguably become an important vehicle on which the Army depends to achieve its mission goals.

The formal pursuit of neuroscience is a theoretical endeavor on the one hand and a practical area of application on the other. Imaging technologies that form the basis for advances in neuroscience have their roots in the medical arena, and it is the Army Medical Research and Materiel Command that has traditionally sponsored much of the basic neuroscience research of benefit to individuals in all military services. DOD-level recognition of the importance of neuroscience research can be seen in the 2008 establishment of the Defense Centers of Excellence for Psychological Health and Traumatic Brain Injury under the assistant secretary of defense for health affairs.

The Army Research Institute for the Behavioral and Social Sciences (ARI) has traditionally conducted research in support of personnel testing and assessment, but neither ARI nor the Army's main research arm, the Army Research Laboratory (ARL), possesses in-house facilities to perform basic neuroscience research. In light of a growing awareness of neuroscience potential in military applications, however, the ARL is planning to establish a collaborative technology alliance on cognition and neuroergonomics, which will take a multidimensional approach (e.g., genetics, computational modeling, neuroimaging, and performance) to optimizing information transfer between the system and the soldier, identifying mental states and individual differences that impact mission-relevant decision making, and developing technologies for individualized analyses of neurally based processing in operational environments.[6]

Neuroscience advances have already led to a broad array of commercial applications and sparked centers for neuroscience research at academic institutions throughout the country and the world. Table 2-1 lists sample objectives in important Army application areas likely to benefit from neuroscience advances. To respond to the statement of task,

[6]Kaleb McDowell, U.S. Army Research Laboratory, "ARL Research in Neuroscience," presentation to the committee, December 18, 2007.

TABLE 2-1 Prospective Army Applications for Neuroscience

Application Areas	Sample Objectives
Training and learning	
Training paradigms and methods	Shorten training cycles; assess training effectiveness
Performance assessments of individuals and groups	Detect individual performance degradation; assess group–individual interactions
Identification of training candidates	Improve success rates
Training effectiveness measures	Predict optimal performance; anticipate degraded performance
Optimizing decision making	
Individual and unit readiness	Utilize neural-state indicators
Adversary assessment and prediction	Act inside adversary decision cycle; disrupt adversary decision making (psychological operations)
Setting objectives	Reduce risk by matching goals with performance
Sustaining soldier performance	
Recovery and reset	Mitigate effect of sleep deprivation on recovery; neuropharmacological intervention to mitigate trauma response
Counterstress	Insulate immune system; moderate disease; modify brain functions to contend with combat rigors
Fatigue and pain	Nutritional countermeasures; minimize effects of sleep deprivation; drug therapies
Brain injury	Intervene early to mitigate acute and long-term deficits due to trauma
Improving cognitive and behavioral performance	
Soldier skills	Optimize brain–machine interfaces; improve image interpretation capabilities
Information utilization and management	Personalize data fusion; prevent information overload

the applications are organized in four categories: training and learning, optimizing decision making, sustaining soldier performance, and improving cognitive and behavioral performance. There is no question that neuroscience research has great potential for the Army's future, but there are societal issues, including ethical considerations and cultural impediments, that must be overcome to realize its full potential.

SOCIETAL ISSUES

The decoding of the human genome and the emergence of new imaging modalities are making possible the identification of proteomic, genomic, and imaging biomarkers associated with susceptibility to a specific disease, with environmental stressors, or with neuropsychological vulnerabilities (e.g., pain, reduced perception, anxiety). The aggregate of multiple biomarkers may provide a susceptibility profile that would not be achievable through testing for any single marker alone. These aggregate data can help in monitoring the rate of progression of clinical disorders or response to treatment. Creating a susceptibility profile of such signatures for a patient can allow for personalized medicine tailored to individual need. The nature (e.g., mutation sites, triplet repeats, proteomic signatures, and structural and functional imaging changes) and quantity of biomarkers involved could play an important role in screening for, diagnosing, and predicting disease. This same capability has made it possible to select persons with a low risk of developing disease or succumbing to a variety of stressors (toxic materials in the environment, for instance). There may be adverse economic consequences (uninsurability, reduced rates of compensation) for the individual and his or her career path progression (costly training programs may intentionally preclude high-risk persons from participation) associated with the identification of disease potential and susceptibility to stress, and so the downside of such information has become of social concern.

Ethical Considerations

One consequence of the genetic screening of large numbers of healthy persons for susceptibility to treatable or manageable disease is that subsequent studies may reveal that the same gene predisposes to an untreatable disease. A case in point is the screening of individuals for a particular allele of apolipoprotein E4, which was known in the 1980s to be associated with high risk for cardiovascular disorder. In the early 1990s it was found that an apolipoprotein E4 also was associated with higher risk for Alzheimer's disease (Corder et al., 1993). Patients who wanted to reduce the risk of cardiovascular disease and signed a consent form for such analysis became aware of their increased susceptibility to Alzheimer's, a then untreatable neurological disorder. Such information was not wanted and caused distress for a significant number of people.

Another consequence of learning one's susceptibility to a nontreatable disease is the still-healthy patient's inability to anticipate the effect of such information on lifestyle and quality of life. Huntington's disease is a clear example of this effect. In the early part of the 1980s it became possible to screen patients and determine whether they would develop the neurodegenerative disorder Huntington's disease, a dementia that does not appear until the fourth decade of life or later and that is associated with an extensive triplet repeat of cytosine-adenine-guanine (CAG) (Myers et al., 1993). Because the disease is autosomal dominant (all individuals with the gene will develop Huntington's disease), it was proposed that all conceptuses that had a parent or close relative with Huntington's disease should be tested for the gene. Knowledge of whether one has a genetic predisposition and hence was certain to develop the disease could have a neutral, beneficial, or adverse effect on the patient, with outcomes ranging from acceptance to suicide. Undesired information on susceptibility to an untreatable disease might be a sword of Damocles for a young soldier.

The positive aspects of knowing how one's genetic heritage impacts wellness and resilience to toxic insult can be illustrated by the case of glutathione S-transferase genotype (GST) and resistance of smokers to lung cancer. A significant number of studies reveal a twofold increase in the incidence of squamous cell carcinoma for patients having GST M1 and GST T1 null genotypes. GST and cytochrome P450 are two classes of enzymes that metabolize and detoxify potential environmental toxins. Patients with the null forms do not express active GST with the properties of GST M1 or GST T1, and the absence of these enzyme variants might predispose them to the toxic effects of various chemicals, including some chemical warfare agents or toxic industrial materials. Restricting the assignment of certain soldiers to areas of high risk might protect them from exposure to such toxic materials; however, it might also keep her/him from serving or being advanced (promoted) in an area of specific interest, or it might prevent him/her from participating in important missions before any clinical manifestation of illness.

Federal laws and regulations contained in the Health Insurance Portability and Accountability Act (HIPAA) protect patients and the community at large against the unwarranted and unnecessary disclosure of medical information that is directly or indirectly traceable to a particular individual to unauthorized parties. The primary concerns are that such disclosure might (1) affect the promotion of military or civilian persons in their field of specialization, (2) affect insurance rates and insurability of a given individual, or (3) affect the psychological/social well-being of individuals with catastrophic diseases that are currently largely untreatable (e.g. Huntington's disease, Alzheimer's disease). These are but a few of the unintended consequences of inappropriate disclosure.

For additional information on ethical issues relevant to neuroscience research discussion in this report, the committee recommends the following sources: Karanasiou et al., 2008; Fins and Shapiro, 2007; Illes, 2007a and 2007b.

Cultural Impediments

The emergence of new bio- and neurotechnologies permits categorizing the human population into subsets having either increased or decreased susceptibility to disease and stress on the basis of their genotype and their phenotype.[7] There is tension between the idea of "selecting out" individuals for tasks based on presumed genetic susceptibilities and the belief that extensive training can overcome inherited limitations and liabilities. The science-fiction film *Gattaca* (Columbia Pictures Corporation, 1997) confronts this dilemma with a dark view of the preselection concept. The film was made during the early stages of the Human Genome Project and the first cloning of large animals.

The U.S. military community traditionally aspires to select individuals for particular tasks or promote them based on excellence during training and performance in the field. Selecting in for extended service in closed platforms such as submarines is rigorous: The training periods are long and there are particular social/psychological requirements. Despite this, very little research has been done on selecting in individuals who have a particular aptitude as assessed by genetic and phenotype testing for a particular military position or job. The decoding of the human genome and the advent of real-time imaging of neural information flow and noninvasive tracing of major fiber pathways provide an opportunity to learn how we can use these novel methodologies to enhance training and personalize it to meet the needs of the soldier, to identify characteristics that are particularly well suited to complex and extreme environments, and to detect the early appearance of uncompensated responses to stress and emerging TBI and post-traumatic stress disorder (PTSD).

The issues confronting the Army include training to criterion (90 percent or better appropriate response to challenge), increasing data flow from deployed aerial and ground sensors, human intelligence, electronic communications, and tempo of engagement with increasing capability of lethality. The need to reduce casualties during force-on-force engagement drives the development of means for conducting combat at large standoff distances and acquiring extensive awareness of the adversary's deployment and capability. At the same time, there is a perceived need to minimize noncombatant casualties, which militates against extensive standoffs. These challenges call for a strategy that allows human cognitive capabilities to operate for 18-20 hours per day, 7 days a week for 12 to 15 months at a high tempo of operations. The most affected group will be the command organization, which is permitted little or no respite from high-tempo decision making and little organized sleep.

The Use and Abuse of Socially Sensitive Demographic Categories as Indicators of Neural State and Performance Capability

As the preceding discussions on societal issues suggest, the committee supports and encourages scientifically validated neuroscience applications across the Army-relevant areas highlighted in Table 2-1 and addressed in detail throughout the report. Ethical considerations, such as those related to genetic screening or improper disclosure of

[7]Phenotype is the result of genes plus environment, and epigenetic changes that occur in the individual after conception may play an important role.

personal medical information, are important constraints that need to be considered, even when the science is adequate for a potential application. In addition, there are a number of issues that require further consideration before being pursued for possible Army application.

One of these issues—the use of performance-enhancing pharmacological agents—was central to the committee's decision to distinguish between uses of neuroscience-related countermeasures (including pharmacological agents) to ameliorate a deficit in performance due to a stressor (Chapter 5) and uses of pharmacological agents to enhance performance beyond an individual's baseline capability (Chapter 6). Cautions and caveats appropriate for these two contexts of application are included in both chapters.

A second major area in which caution must be exercised when considering application to Army-relevant problems concerns statistically significant differences in group-averaged neural states or activity patterns—or even differences in behavior—between a demographically defined subpopulation and a reference population. In the case of gender, such differences are typically expressed as a comparison of male and female subpopulations. For other subpopulations of societal interest, such as ethnocultural identity, age, or socioeconomic status, comparisons may be drawn either between one such category (e.g., African-Americans, young people between 18 and 25) and the general population or among subpopulation categories within a classification (e.g., comparisons among ethnocultural groups or age groups).

In most cases, these differences, even when statistically significant, represent differences in population distributions where the distributions have substantial overlap. From an epidemiological perspective, the statistically significant difference in the distributions justifies identifying membership in certain subpopulation categories as a differential risk factor. However, for purposes such as selecting, assigning, or qualifying individuals for a task, the overlap in the subpopulation distributions means that these categories lack the sensitivity and specificity to be reliable indicators of the neural state or behavior of interest. Rather than relying on a familiar but scientifically indefensible population category as a criterion, the appropriate use of neuroscience insights is to seek out one or more truly reliable indicators for the variable of interest.

As a simple but germane example, a number of behavioral epidemiologic studies have found that women are at greater risk for developing PTSD, given similar stress experiences, than men (e.g., Breslau and Anthony, 2007; Turner et al. 2007). Should this difference in relative risk be used, for example, to exclude women from high-stress combat situations? The committee's position is that gender is not a sufficiently reliable indicator of the PTSD outcome to be used as a criterion in selecting and assigning individual women, even though the studies establish being female as a risk factor for PTSD. The numbers of false positives and false negatives are too high; the correlation lacks sensitivity and specificity.

The work by Ursano et al. (2008) illustrates how the neurosciences can extend and inform behavior-based findings—such as the PTSD studies cited above—by opening the way to reliable indicators. Their work indicates that the 5-HT2A receptor, p11 protein, and associated regulators may play a role in PTSD-related response to stress experiences. If this still-preliminary line of inquiry were to lead to suitably sensitive and specific indicators of PTSD susceptibility, then those indicators would be candidate criteria for selection and assignment decisions, where gender is not. In short, whether one is female or male is not the issue; it is one's neurophysiological sensitivity to a definable level and type of environmental stress for which the Army needs validated, reliable indicators.

Over the past half-century and longer, American society has traveled a long and difficult road to break away from unscientific stereotypes about gender, "race," and other previously accepted ways of categorizing individuals to define their suitability for various roles and responsibilities. The committee's deliberations on how to deal with research on gender differences and other demographic categories acknowledged these societal issues. In light of its concerns about applying group-averaged statistical differences to individuals within a group, the committee decided to emphasize individual variability in neural-based traits and tendencies as a more appropriate way to address observed distributions in a population of interest. By focusing attention on individual variability and the search for reliable indicators of that variability, the committee hopes to avoid unscientific and unethical application of findings about behavior and neurophysiology in ways that would negate our hard-won progress toward fair treatment and equal opportunity.

REFERENCES

Aston-Jones, G., and J.D. Cohen. 2005a. Adaptive gain and the role of the locus coeruleus-norepinephrine system in optimal performance. Journal of Comparative Neurology 493(1): 99-110.

Aston-Jones, G., and J.D. Cohen. 2005b. An integrative theory of locus coeruleus-norepinephrine function: Adaptive gain and optimal performance. Annual Review of Neuroscience 28: 403-450.

Atkinson, A.J., Jr., W.A. Colburn, V.G. DeGruttola, D.L. DeMets, G.J. Downing, D.F. Hoth, J.A. Oates, C.C. Peck, R.T. Schooley, B.A. Spilker, J. Woodcock, and S.L. Zeger. 2001. Biomarkers and surrogate endpoints: Preferred definitions and conceptual framework. Clinical Pharmacology & Therapeutics 69(3): 89-95.

Bakker, A., C.B. Kirwan, M. Miller, and C.E.L. Stark. 2008. Pattern separation in the human hippocampal CA3 and dentate gyrus. Science 319(5870): 1640-1642.

Belenky, G., N.J. Wesensten, D.R. Thorne, M.L. Thomas, H.C. Sing, D.P. Redmond, M.B. Russo, and T.J. Balkin. 2003. Patterns of performance degradation and restoration during sleep restriction and subsequent recovery: A sleep dose-response study. Journal of Sleep Research 12(1): 1-12.

Breslau, N., and J.C. Anthony. 2007. Gender differences in the sensitivity to posttraumatic stress disorder: An epidemiological study of urban young adults. Journal of Abnormal Psychology 116(3): 607-611.

Corder, E.H., A.M. Saunders, W.J. Strittmatter, D.E. Schmechel, P.C. Gaskell, G.W. Small, A.D. Roses, J.L. Haines, and M.A. Pericak-Vance. 1993. Gene dose of apolipoprotein E type 4 allele and the risk of Alzheimer's disease in late onset families. Science 261(5123): 921-923.

Fins, J.J., and Z.E. Shapiro. 2007. Neuroimaging and neuroethics: Clinical and policy considerations. Current Opinion in Neurology 20(6): 650-654.

Friston, K.J., E. Zarahn, O. Josephs, R.N.A. Henson, and A.M. Dale. 1999. Stochastic designs in event-related fMRI. NeuroImage 10(5): 607-619.

Genik, R.J., C.C. Green, F.X. Graydon, and R.E. Armstrong. 2005. Cognitive avionics and watching spaceflight crews think: Generation-after-next research tools in functional neuroimaging. Aviation, Space, and Environmental Medicine 76(Supplement 1): B208-B212.

Illes, J. 2007a. Empirical neuroethics. Can brain imaging visualize human thought? Why is neuroethics interested in such a possibility? EMBO Reports 8(S1): S57-S60.

Illes, J. 2007b. Neuroethics in a new era of neuroimaging. Pp. 99-103 in Defining Right and Wrong in Brain Science: Essential Readings in Neuroethics. W. Gannon, ed. Washington, D.C.: Dana Press.

Karanasiou, I.S., C.G. Biniaris, and A.J. Marsh. 2008. Ethical issues of brain functional imaging: Reading your mind. Studies in Health Technology and Informatics 137: 310-320.

Kurzweil, R. 2005. The Singularity Is Near. New York, N.Y.: Viking Press.

Myers, R.H., M.E. MacDonald, W.J. Koroshetz, M.P. Duyao, C.M. Ambrose, S.A.M. Taylor, G. Barnes, J. Srinidhi, C.S. Lin, W.L. Whaley, A.M. Lazzarini, M. Schwarz, G. Wolff, E.D. Bird, J.-P.G. Vonsattel, and J.F. Gusella. 1993. *De novo* expansion of a $(CAG)_n$ repeat in sporadic Huntington's disease. Nature Genetics 5(2): 168-173.

Otten, L.J., R.N.A. Henson, and M.D. Rugg. 2002. State-related and item-related neural correlates of successful memory encoding. Nature Neuroscience 5(12): 1339-1344.

Penfield, W., and P. Perot. 1963. The brain's record of auditory and visual experience: A final discussion. Brain 86(Part 4): 595-696.

Turner, J.B., N.A. Turse, and B.P. Dohrenwend. 2007. Circumstances of service and gender differences in war-related PTSD: Findings from the National Vietnam Veteran Readjustment Study. Journal of Traumatic Stress 20(4): 643-649.

Ursano, R.J., H. Li, L. Zhang, C.J. Hough, C.S. Fullerton, D.M. Benedek, T.A. Grieger, and H.C. Holloway. 2008. Models of PTSD and traumatic stress: The importance of research "from bedside to bench to bedside." Progress in Brain Research 167: 203-215.

3

Training and Learning

This chapter discusses opportunities to expand on and improve the Army's current behavioral approaches to training and learning by incorporating recent or emerging advances in neuroscience. The discussion is organized under five application areas:

- Evaluating the efficiency of training regimes and learning paradigms,
- Individual capability and response to training,
- Monitoring and predicting changes in individual performance efficiency,
- Soldier selection and assessment, and
- Monitoring and predicting social and group interactions.

The final section summarizes applications of neuroscience to Army training and learning in terms of when practical application can be expected: 5 years (near term), 10 years (medium term), or 20 years (far term). Enhancing the utility and predictive power of traditional behavioral and psychological methods by incorporating the insights and tools of neurophysiological monitoring in these and other application areas—but focused on understanding how individual choices are made in nonmilitary contexts—is a principal goal of neuroeconomics.[1] Results from neuroeconomics research are cited frequently in this chapter and Chapter 4.

EVALUATING THE EFFICIENCY OF TRAINING REGIMES AND LEARNING PARADIGMS

Neuroscience offers new ways to assess how well current training paradigms and accepted assumptions about learning achieve their objectives. For example, does overtraining in specific paradigm responses increase trainees' agility in responding to new threats, or does training in probabilistic assessment make them more adaptable? Two examples illustrate how each methodology produces benefits in specific cases. From a behavioral perspective, overtraining reduces time to response by reducing consideration of alternatives in favor of an anticipated favorable outcome to the trained-in response (Geva et al., 1996). By contrast, a medical doctor is trained to diagnose any given set of clinical presentations (e.g., presenting symptoms, past individual and family medical history) in terms of a probabilistic etiology based on environment, current illnesses in the community, and other factors. This differential diagnosis strategy helps focus possible treatment modes on likelihood of outcomes. When sufficient evidence accrues, the most likely cause may not be the most frequent cause.

Neuroscience-Based Models of Learning

Over the past decade, enormous progress has been made toward describing the neurological basis for learning skills and procedures. We now have fairly complete models that describe how the brain learns the values of actions (see, for example, Niv and Montague, 2009) and uses these values to guide future decisions, a process often called reinforcement learning. Suppose a subject is offered the opportunity to search in one of two locations for a reward on each of hundreds of sequential trials. With repeated sampling, the subject learns the relative values of the two locations and shapes his behavior to maximize his reward. Although human subjects show idiosyncratic behavior under this training regime, the learning models are now well enough developed that an individual's behavior can be well characterized by a single parameter. Given the value of this parameter for an individual, his future choices can be predicted with accuracies approaching 90 percent (Corrado et al., in press). These learning models thus both describe the behavior observed in humans and animals and predict how a subject, once char-

[1]Neuroeconomics is characterized in a current survey of the field as the convergence of normative models of choice—the province of economics—with the psychological and neurobiological processes (or algorithms) by which individuals (animal and/or human) make choices (Glimcher, in press).

acterized, will behave in dynamic environments (Balleine et al., 2009).

Similarly, tremendous advances have been made in understanding movement generation, including development of skills, habits, and automatic performance (Poldrack et al., 2005; Yin and Knowlton, 2006). Moreover, consider an individual who is learning to execute a complex movement accurately—for example, tracking a moving target with his finger (Poulton, 1974). We understand quite precisely how the incentives and feedback provided shape both performance and learning (Newell, 1996; Trommershäuser et al., 2009). We also have evidence that the fastest way to train a movement is not to provide strictly accurate feedback. The speed and effectiveness of such training can often be maximized by providing feedback regimens that take advantage of this inherent accommodation to variability (Schmidt and Lee, 2005; Kording et al., 2007).

In short, advanced models of reinforcement learning and movement control learning have implications for both training and prediction of learning efficiency. We know in principle how to develop optimal training regimes under many conditions, and we know how to predict agent behavior with precision under a range of conditions. Both assets could be leveraged to improve not only training but also data presentation (for situational awareness) and prediction of threat/enemy behavior.

The combination of neuroimaging tools, cognitive neuroscience, and experimental or cognitive psychology has resulted in the development of models of how the brain may process information. For example, recent accounts of brain processing that occurs in the dorsolateral prefrontal cortex (PFC) are based on interpreting cognitive control as altering ongoing behaviors in order to adjust to the changing context of the environment (Botvinick et al., 2001). The resulting computational models reveal which aspects of cognitive performance are altered as information changes—a field of study called "computational neuroscience"—and can be used to predict performance on behavioral tasks (Brown and Braver, 2007).

Computational neuroscience uses mathematical models to study how neural systems represent and transmit information. The discipline may be roughly divided into two schools. The first school uses detailed biophysical models of individual neurons, detailed models of neuronal networks, and artificial neural network models to study emergent behaviors of neural systems. The second school develops signal-processing algorithms, computational models, to analyze the growing volumes of data collected in neuroscience experiments. In these computational models, adaptation to new information is represented as changes in one or more parameters. By combining the models of both schools with information from neuroimaging tools (sometimes called "systems neuroscience") and behavioral neuroscience, the possible causes of underperformance and the conditions conducive to improved performance can be quantitatively constrained. Eventually, it may be possible to develop specialized interventions aimed at improving performance in soldiers.

Subject Populations for Army-Specific Studies of Learning and Training

A perennial issue for behavioral and neurological testing is the degree to which experimental findings from a specific sample can be extrapolated to a target population. Many of the activities in which soldiers and their leaders engage depend critically on rigorous preparatory training and task-specific expertise. The majority of behavioral research is performed with subjects who are either patients in a clinical setting or volunteers from a university community (mostly undergraduate students). A critical question is how far results based on these study populations of convenience transfer to a soldier population. In cases where the research hypothesis addresses Army-relevant issues directly, the typical research subject populations may not be sufficiently representative of the population to which the Army wants the results to apply. In short, neither clinical patients nor university undergraduates are good surrogates for a soldier.

As the Army seeks to apply research results from the neuroscientific literature, the extent to which they can be transferred to a military (specifically, Army) population should itself be a subject of Army research. In particular, how far do results from typical civilian samples represent those to be expected from an Army population? Although some of the research reviewed by the committee has used actual soldiers or cadets, for the most part the human subjects in potentially relevant studies do not compare well to the soldier population in cardiovascular fitness, psychological drive to perform, and learning/training experiences that clearly affect neurobehavioral response—e.g., boot camp, intense training for operational performance, and actual operations.

One alternative to constraining Army-usable results to just the few studies that use soldiers (or even military cadets) is to seek subject populations that more closely resemble Army soldiers in such key characteristics as cardiovascular fitness, psychological motivation to perform, and training/learning in immersive, demanding environments. High-performance athletes are one such subpopulation, and there is an extensive literature of behavioral and neuropsychological research on them. Appendix C lists a sampling of the research literature from 2001 through 2007 on training methods for high-performance athletes: performance evaluation/assessment of athletes in training, including under stress; social interaction with other athletes; and issues with performance anxiety and other psychological issues including depression in ex-athletes (references 1-109 in Appendix C). Several studies have investigated the use of mental imagery in training athletes and its effects on performance (references 110-123). Performance after mild concussions and determining when the subject can return to a normal (strenuous) routine is a hot topic for both athletes and soldiers, and some longer-term studies investigating the effects on athletes of multiple concussions have also been published (references 124-143). Studies of training and performance issues for female athletes go well beyond the well-researched Female Athlete Triad[2] (references 144-151). The relationship between athletes' risk-taking behavior and athletic performance is the topic studied in references 152-157. Other relevant topics in this literature include the relationship between the lifestyle of athletes and changes in immune function (references 158-160), and the effect of music therapy on athletic performance (reference 161) and of caffeine (reference 162).

While it is important to consider how well the subject populations in the research studies match the population of interest (here, Army soldiers), the context of the research must also be considered. For example, in testing of futuristic military decision support systems, military subject-matter experts perform worse than novice populations, owing to cultural biases (Graham et al., 2007). When designing a futuristic system, it may therefore be better to test it using a subject population likely to possess the required skills (including constructive attitudes toward novel representations), which current military personnel may lack.

INDIVIDUAL CAPABILITY AND RESPONSE TO TRAINING

Given the increasing technology- and threat-driven dynamic complexity on the battlefield, it has become more and more critical to optimize the capability of individual soldiers. One solution to the great differences from one individual to the next in human capability and expertise is to tailor human–system interfaces to human capabilities and to adapt training regimes to the individual. Although classic experimental psychologists tended to downplay individual differences in their theories of human cognition, educators have consistently reminded psychologists of the need to understand individual differences in cognition and performance (Mayer, 2003). A similar need exists in understanding how the neural systems underlying cognition differ among individuals (Posner and Rothbart, 2005). Understanding the neural substrates of individual differences in cognition can help in characterizing the differences, developing training methods tailored to them, matching individuals to assignments for which they are well suited, and optimizing human–machine and individual–system interfaces.

Recent advances in neuroimaging make this endeavor possible (Miller et al., 2002; Miller and Van Horn, 2007). Patterns of brain activity as measured by functional magnetic resonance imaging (fMRI) appear to provide unique identifying characteristics. These are unlike fingerprints, because individual brain activity may not be epiphenomenal in the sense of simply being a developmental outcome that does not causally influence individual behavior. Rather, individual patterns of brain activity may reflect (or underlie) the unique characteristics of individual minds, and they may capture aspects of an individual mind that cannot be obtained using conventional behavioral measures or self-reporting. The better we understand the sources of individual variability in brain activity through systematic experimentation and analysis, the more fully we can determine whether, and to what extent, these individual differences can be used to assess and ultimately train that person.

Individual Variability in Brain Activity

Efforts by neurologists and neuropsychologists to understand individual differences in brain processes go back to at least the mid-1800s, when the neurologist Paul Broca concluded, from examining the common area of damage across a group of patients exhibiting similar speech production deficits, that speech production could be localized to the third convolution of the left inferior frontal gyrus. Around the same time, the neurologist John Hughlings Jackson argued against a centralized region for speech, basing his opinion on his observations of wide variations in the extent and location of damage in patients exhibiting similar problems in speech perception and wide variations in symptoms in patients with similar damage. More than a century later, Broca's view of brain organization became the dominant paradigm. An example is the Wernicke-Geschwind model, which was developed to explain language function. Today, however, the Broca paradigm is often disregarded, largely because of enormous individual variability in the underlying brain processes. This variability is evidenced by the fact that a growing number of neurosurgeons painstakingly map out individual brains just prior to surgery, after a portion of the patient's skull has been removed but while the patient is still conscious and responsive.

Researchers using neuroimaging to understand the relationship between the brain and the mind have recently encountered a similar paradigm shift. Most neuroimaging studies localize cognitive functions in the brain by conducting a statistical analysis across a group of subjects; this analysis identifies common areas of activation. While this can be a useful approach to understanding the modular organization of the brain, it disregards the not-common areas of activation that can be observed at the individual level and that may also be critical for that function in a given individual. Recent studies have shown that the individual patterns of brain activity during a memory task are enormously variable, sometimes with areas of activation that do not even overlap between two subjects (Miller et al., 2002; Miller and Van Horn, 2007). Furthermore, such studies have found that individual variations in brain activity could not be attributed to random noise because the pattern for an individual is stable

[2]This is a condition of three syndromes common in high-performing female athletes of all ages, though especially in their teen years, and includes disordered eating, amenorrhea, and osteoporosis. An athlete can experience one, two, or all three syndromes in the triad.

over time. Understanding these individual differences in brain activity has not only significant theoretical implications but also pragmatic implications for the Army in trying to understand how individuals respond to and remember events in battlefield and nonbattlefield situations and in characterizing the mental traits of individual soldiers and officers.

Individual variability in brain activity could come from one or two things that are not mutually exclusive. One possibility is that individual differences in brain anatomy and physiology lead to extensive differences in brain activity (for example, as measured by the BOLD signal in fMRI) despite the tremendous efforts undertaken to normalize brain spatial variability to a standard spatial representation. Structural differences in brain anatomy and physiology take many forms. It is known, for instance, that the size and shape of individual brains can vary greatly, and genetic markers associated with this variability have been found (Tisserand et al., 2004). There are also well-documented individual differences in the tissue structure of specific regions of the cerebral hemispheres (cytoarchitectonics) and in the orientation and location of specific fissures and gyri (Rajkowska and Goldman-Rakic, 1995). Theoretically, these regions are spatially normalized in fMRI analyses using sophisticated algorithms so that brain regions are consistent across subjects, but some differences have been shown to exist even after extensive spatial normalization. Other structural differences may also exist, despite the fact that brain development is generally universal (Kosik, 2003; Rakic, 2005).

How anatomical differences affect cognition is not well understood, but there are intriguing possibilities. For example, the size and location of the planum temporale is thought to affect language processing (Hutsler et al., 1998). Other recent studies have linked individual differences in white-matter connectivity to individual differences in cognition (Baird et al., 2005; Ben-Shachar et al., 2007). Understanding the extent to which structural differences account for variability among individuals in brain activation will greatly enhance our knowledge and understanding of individual minds, allowing human–system interfaces to be better tailored.

The second possible source of variability in brain activity is individual differences in cognitive styles, abilities, and strategies. These differences, which appear important to how individuals perform many cognitive tasks, often can be correlated with significant differences in BOLD activity, despite procedural efforts to constrain and control the psychological state when test subjects are performing the same experimental task. For example, a recent study found that brain activations underlying a standard memory task were extremely variable from subject to subject. These differences extended well beyond spatial normalization or relatively small differences in the location of brain structures (Miller et al., 2002). A significant portion of this variability correlates with differences in retrieval strategies (Donovan et al., 2007).

Episodic memory, which relies on an extensive hippocampal-cortical network for the consolidation, storage, and utilization of information, provides a useful model for understanding how individual differences in cognitive style and strategy may affect patterns of activations (Tulving, 2002; Squire et al., 2004). A hypothesis supported by patient studies and animal models is that the hippocampus is not involved in the permanent storage of information per se but rather serves to facilitate consolidation of a distributed cortical memory trace. A principal characteristic of this distributed network is that it allows the rapid and flexible formation of multimodal memories. In this emerging picture of brain activity, episodic retrieval makes use of several distinct brain regions, some of which may be involved in the cognitive processing that is peripheral to the actual retrieval of stored information. The same individual may, at particular times, differentially engage those brain regions based on the context and strategy of the moment. One potential implication of this architecture is that the same behavioral outcome—such as an "old" response on a recognition test—could be based on distinct sets of information and distinct combinations of neural circuits in two different subjects. The recent development of new neuroimaging techniques that enable the systematic study of individual differences in brain activity can be used to improve understanding of the optimal brain activity that underlies, for example, optimal performance on the battlefield or in other demanding environments and task contexts.

To improve the assessment of individual soldiers, research programs should systematically investigate the structural and cognitive factors that may account for the extensive individual variability that has been observed in brain activity across normal subjects using fMRI. Three basic questions need to be answered:

- Why do individuals' patterns of brain activity differ so much from each other?
- Can individual differences in brain activity be accounted for by differences in anatomy or physiology? Can information about individual brain activity be indicative of limitations and constraints in the kinds of cognitive strategies and skills that a particular individual is capable of or tends to engage in?
- Can individual differences in brain activity be accounted for by differences in cognitive style and strategy? Can information about individual brain activity be used to assess the thought processes of individuals engaged in a variety of tasks?

Identifying Conceptual Change in Individual Learning

The potential utility for the Army of monitoring individual differences in brain activity to track and evaluate individual learning can be illustrated by recent work in correlating differences in brain activity with success in assimilating a basic concept of physics that is nonintuitive. Teachers and researchers have found that beginning physics students retain a naive "impetus" theory of motion that differs from the fundamental concepts of force and motion central to Newtonian physics. Many physics courses may be needed before the Newtonian concepts become a student's "natural" way of seeing the world (Clement, 1982; McCloskey, 1983; Mestre, 1991).

A number of studies have examined the degree to which major changes (through either education or experience) in the way people view the world can be captured using fMRI. Recently, fMRI has been used to compare the brain activity patterns of students who had taken no high school or college physics with the patterns of students who had taken at least five college-level physics courses (Dunbar et al., 2007). Students were shown similar movies of two balls falling at either different rates or at the same rate. The students were asked whether the movie they viewed was consistent or not with their expectations. The researchers were particularly interested in movies where balls of different sizes (and therefore different apparent mass) either fell at different rates (as expected in an "impetus physics" view) or fell at the same rate (as expected in a "Newtonian physics" view). The fMRI data showed that a particular brain region associated with error detection, the anterior cingulate cortex, was activated in the nonphysics students when they saw the balls of different sizes drop at the same rate, while the same region was activated in physics students when they saw the balls drop at different rates. Conversely, neither group showed the characteristic error detection activity when the movie they saw fit the expectations of their engrained physics concepts. The researchers concluded that the physics students had indeed fully assimilated the Newtonian concepts. In the context of Army applications of neuroscience, this example illustrates how fMRI could be used as an indicator of whether soldiers have learned and assimilated key concepts fully enough to act on them instinctively.

MONITORING AND PREDICTING CHANGES IN INDIVIDUAL PERFORMANCE EFFICIENCY

This section begins by outlining some conditions in which neuroscience approaches may be useful in assessing changes in the performance efficiency of individual soldiers that are due to the stresses of extreme environments and combat operations. Advantages and disadvantages of potential neuroscience approaches are illustrated with a few examples. Finally, short-, medium-, and long-term opportunities for Army R&D on these approaches are discussed.

The Effects of Environmental Stressors on Individual Performance

Extreme environments, including but not limited to combat environments, are characterized by the high demand they place on physiological, affective, cognitive, and social processing in the individuals exposed to them. In short, the stressors present in extreme environments strongly perturb the body and mind, which in response initiate complex cognitive and affective coping strategies. Research on expedition members, soldiers, elite athletes, and competitors in extreme athletic events provides substantial evidence that exposure to extreme situations profoundly affects performance. Different environmental stressors can place different demands on individuals or groups exposed to them. For example, exposure to extreme cold during an Antarctic expedition may result in social deprivation, whereas exposure to combat may result in affective overload. However, beyond these differences in response that correspond to differences in the environmental stressors, individual cognitive and affective responses to the same stressor vary just as widely.

Neuroscience offers some distinct advantages over behavioral assessment or self-reporting for assessing and even predicting how an individual's baseline performance is affected prior to, during, and following exposure to a particular environmental stressor. One key limitation of standard self-reports and observer reports is their limited ability to predict future behaviors. Moreover, a number of studies have shown that individuals do not always report their current psychological, mental, or emotional status accurately (Zak, 2004). Although there is still no single neural measurement tool that can unequivocally replace self-reporting, the current tools can depict an individual's status more fully, as a complement to, rather than a replacement for, self-reporting and behavioral assessments. Moreover, recent insights enable researchers to parse cognitive and emotional processes into more basic modules such as attention, working memory, cognitive control, and others. These modules can be assessed efficiently and quantitatively by linking behavioral paradigms to measurements made with electroencephalography, fMRI, or other imaging modalities.

Finally, behavioral tasks that have been developed recently for use with these imaging modalities are parametric in the sense that the imaging results can be used to quantify the degree to which performance is altered. Quantifying the degree of underperformance (the performance deficit relative to the individual's baseline) is crucial to designing and administering countermeasures. (Chapter 5 discusses in more detail some countermeasures to stressors that degrade soldiers' cognitive performance.)

A major challenge for neuroimaging used in this way is to determine its sensitivity and specificity for monitoring performance in extreme environments.[3] Thus far, most

[3]As discussed in Chapter 2, sensitivity and specificity have rigorous definitions that apply here. Thus, "sensitivity" measures the proportion of the actual positive cases that a test identifies as positive. Mathematically, it is the ratio of true positive test results to the sum of the true positive tests and the false negative tests (false negatives should have been positive). "Specificity" measures the proportion of actual negative cases that the test identifies as negative. It is the ratio of true negative test results to the sum of the true negative tests and the false positive tests (false positives should have been negative). The aim of having a test with both high sensitivity and high specificity is to identify all the positive cases efficiently while also distinguishing positive from negative cases.

imaging studies have revealed intriguing results on a group-averaged level. But a key goal of a neuroscience-based approach to quantifying behavior and performance is greatly improved ability to predict the future behavior of individuals (individual-specific prediction). It is not enough to predict the distribution in performance of a group (or parameters of that distribution such as the average performance or a "normal range"). The goal is to predict which individuals will perform well and which will not, and even why individuals perform as they do. Recent results from neuroimaging studies indicate that such predictions are possible. Most imaging studies have demonstrated large effect sizes, which supports the idea that differences among individuals and at different times for the same individual may be large enough to be meaningfully measured.

The next step is to determine if neuroimaging or other neuroscience approaches that combine sensitivity and high specificity can be used to generate quantitative predictions of individual behavior in response to environmental stressors. To be more useful as a predictor of performance than current approaches, neural monitoring methods such as neuroimaging need to track performance states closely, including considering whether or not the individual is self-reporting a change in performance and whether or not the self-reporting is as accurate as objective measures of performance. To be useful for identifying (and eventually predicting) performance deficits that are due to extreme environments, the neural monitoring methods must do more than distinguish poor-performing individuals from normal performers. They must also consistently distinguish altered activity from normal activity in specific brain structures of those individuals who subjectively report performing at normal levels but whose performance has deteriorated by objective measures. The last-mentioned capability will enable the identification of individuals whose performance in future missions is at high risk of deteriorating, as well as the identification of design factors for military equipment. The example in Box 3-1 illustrates the potential of neuroscience methods to make such predictions.

SOLDIER SELECTION AND ASSESSMENT

Currently the Army selects most of its individual soldiers using two basic tests, both administered by the Department of Defense: the Armed Services Vocational Aptitude Battery (ASVAB), for selecting enlisted personnel, and the Scholastic Aptitude Test (SAT), for selecting officers. These tests are used for assessing both an individual's fitness for the Army and his or her suitability for specific assignments. Neither test assesses personality traits or neuropsychological traits of applicants.

There has been little attention to how the Army could select, from a general pool of applicants, those individuals who, by inherent or acquired ability, would add value to a particular unit. Even for assignments that require expensive and specialized training, the Army knows little about a candidate's neuropsychological traits before that individual starts training. Little is known about how to identify soldiers whose individual traits would enhance the performance of a unit to which they could be assigned. Indeed, for many high-performance assignments, one cannot state with certainty which psychological or behavioral traits correlate with superior performance. For example, the training for an attack helicopter pilot costs about $225,000, but candidates for this training are selected today largely on their expressed interest in becoming an attack helicopter pilot rather than on their ability or fitness for this high-value assignment.

At present, the Army has low washout rates even for

BOX 3-1
Predicting Future Behavior in Extreme Environments

Imaging techniques could be used to detect individuals who are at high risk for experiencing deterioration of performance on future Army missions. Paulus et al. (2005) used fMRI to scan the brains of 40 methamphetamine-addicted men who had been sober for 3 to 4 weeks. This imaging technique can map brain regions involved in specific mental activities. Scans were performed while the men were involved in a decision-making task to identify brain regions stimulated by the task. Approximately 1 year later, the researchers correlated the fMRI results with subsequent drug abuse in the 18 men who relapsed and the 22 who remained abstinent (drug-free).

In the scans made at the beginning of the study, the scientists observed low activation patterns in the brains of some of the men in structures that are known to participate in making decisions. These regions were the right middle frontal gyrus, the right middle temporal gyrus, and the posterior cingulate cortex. Lower activity in these structures correlated with early relapse to methamphetamine use. The scans also showed that reduced activation in the insula and the dorsolateral prefrontal, parietal, and temporal cortices correlated with drug relapse in 94 percent of cases. By comparison, significant activation in these same regions correlated with nonrelapse in the 86 percent of the men who remained abstinent after 1 year. Thus, the initial brain scans provided an indicator of which individuals were at greatest risk of relapsing and which were at least risk. Furthermore, these differences in activation pattern show both specificity and sensitivity as a predictive indicator for risk of relapse.

difficult jobs because trainees are free to repeat the training sequence until they are able to meet minimum performance requirements. Less than 5 percent of helicopter pilot trainees, for example, wash out of training. This approach obviously increases training costs. It may also reduce unit performance; a soldier who requires three times longer than average to achieve minimal standards of proficiency during flight training may or may not be a good helicopter pilot. At the moment, however, the Army has no validated means of measuring how this approach affects unit performance.

Current Enlisted Soldier Selection

The ASVAB, which is administered to roughly 500,000 applicants each year, is a computer-administered multiple-choice test with two components: the Armed Forces Qualification Test (AFQT) and five subject area tests. The goals of the AFQT are to predict the likelihood of service completion by the applicant and to predict overall performance. The AFQT has four parts: word knowledge, paragraph comprehension, arithmetic reasoning, and mathematics knowledge. AFQT scores are percentile rankings that have been normed to the ASVAB applicant pool and are correlated with the likelihood that the applicant will complete a 2-year tour of duty. The higher the score, the more likely the applicant is to complete that tour. Interestingly, the AFQT prediction of tour completion is most accurate for applicants who hold a traditional high school diploma. The AFQT score is a poor predictor of tour completion for applicants who did not graduate from high school or who have a graduate equivalency diploma.

To address the limitations of the AFQT, the Army Research Institute (ARI) has developed an alternative test for candidates who did not complete high school. This "noncognitive" test, called the Assessment of Individual Motivation (AIM), was designed to assess conscientiousness, stress tolerance, and openness to new experiences. The AIM in combination with a measure of body mass index is called the Tier Two Attrition Screen and is now in limited use as a complement to the ASVAB.

The second part of the ASVAB consists of five tests of the applicant's factual knowledge in general science, automobile and shop information, mechanical comprehension, electronics information, and assembling objects. The scores on these tests by soldiers who have successfully completed a tour of duty in a particular area of specialization are used as a benchmark or score profile that is associated with successful tour completion in that area of specialization. The score profiles are used by recruiters and recruits as guides to a recruit's likely job performance when the recruiter and recruit together select an area of specialization. Essentially, the score profiles from the five subject tests are derived by simple linear regression. Their actual predictive power is low compared with that of best practices in the vocational assessment and training community.

Advanced Soldier Selection Tools

For many high-value soldier assignments, the ASVAB may only weakly predict how long it will take to train a recruit, how likely the recruit is to complete a tour of duty, or even whether he or she will be effective in the assignment. The Army recognizes the poor predictive power of the ASVAB and has commissioned ARI to construct secondary selection tests for a number of high-value specialties that call for a substantial investment in training. There are now specialized selection tests for helicopter pilots, improvised explosive device (IED) detection experts, Special Forces, drill sergeants, and recruiters. In general, ARI has developed these specialized selection tests as psychological tests.

By developing and using these tests as selection tests—even if only to a limited extent—the Army has in fact implemented a screening process for those interested in certain high-value assignments. For example, the helicopter pilot test allows some recruits to become helicopter pilots and excludes others from the required training. The question, then, is not whether applicants should be screened, at least for demanding and high-value positions, but rather how well the screening instruments do their job. To date there appears to have been little effort to ensure that these tests have both specificity and sensitivity with respect to predicting actual performance.

To illustrate the state of practice in Army selection testing and highlight the challenges where neuroscience-based approaches can help, the discussion will focus on the Army's current test for helicopter pilots, the Alternate Flight Aptitude Selection Test (AFAST) and the test proposed as its replacement. After taking the ASVAB, Army recruits who want to train as helicopter pilots are invited to take the AFAST, which is a computer-based test. The Army uses the AFAST score to assess the suitability of the recruit for flight training. Two factors limit the effectiveness of the AFAST as a screening tool. First, the security of the AFAST has been compromised; answers to the test can be found on the Internet. Second, according to the briefing ARI gave to the committee, the AFAST has almost no predictive ability.[4]

In recognition of the limitations of the AFAST, ARI was tasked with developing an alternative, the Selection Instrument for Flight Training (SIFT). As an instrument for soldier selection, the SIFT model is based more closely than AFAST on the selection techniques used in private industry, although it is still limited to conventional behavioral testing. The SIFT is a secure test that measures, among other things, a number of personality features, perceptual speed and accuracy, and flexible intelligence. It has almost triple the predictive accuracy of the AFAST for helicopter pilot selection, based on subsequent pilot performance and aptitude evaluations by peers and instructors. The SIFT score has not yet been correlated with in-theater performance of helicopter pilots who

[4]Lawrence Katz, research psychologist, Army Research Institute, briefing to the committee on April 28, 2008.

took it prior to training. Unfortunately, there are no definite plans to gather the data to establish this correlation, a final assessment that is critical for such a tool.

Despite its advantages over the AFAST, the SIFT nevertheless lacks any neuropsychological measurements, nor does it look at neurochemical or genetic indicators of neuropsychological traits. ARI does not have expertise in these neuroscience-based measurements. Enhancing the SIFT by giving it the ability to identify established neuropsychological correlates of the personality and cognitive traits characteristic of highly successful helicopter pilots would greatly improve its value as a screening instrument. For example, research has found that individuals whose hypothalamo-pituitary axes are highly reactive to stress are unlikely to complete Navy SEAL training (Taylor et al., 2006, 2007). However, such information is not yet used by the Army in its recruit assignment process.

Even after advanced selection instruments have been developed, implementing them throughout the Army will be a challenge. Recruiting stations throughout the United States must be able to administer specialized selection tests like the SIFT, whose implementation cost has been estimated at roughly $200,000. Although this implementation cost is less than the cost of training a single helicopter pilot, ARI has not yet succeeded in getting the SIFT implemented generally in the recruitment process.

A similar project is under way to improve the test instrument used for identifying trainees in explosive ordnance disposal. This project, which is conducted jointly by ARI and the Joint Improvised Explosive Device Defeat Organization, was not evaluated by the committee because it involves classified information.

Neuropsychological Testing in the Army: The Automated Neuropsychological Assessment Metrics

The only Army testing instrument that contains significant elements of neuropsychology and cognitive neuroscience is the Automated Neuropsychological Assessment Metrics (ANAM). ANAM evolved from a series of neurological assessment tools initially developed during the Vietnam war. The goal was to develop behavior-based tests for determining whether the cognitive capability of a soldier had been impaired by exposure to a chemical agent and for testing protective agents for side effects. A group of such tests was consolidated by a multiservice group into what later become the ANAM. In its current form, ANAM is a computerized neuropsychological battery of 30 test sets, some of which have not been normed. ANAM support and development are currently under the direction of the Center for the Study of Human Operator Performance at the University of Oklahoma.

An ANAM assessment ranges from simple tests of reaction time to dual-task interference tests, which are useful for assessing executive function. The test sets are essentially standard neuropsychological tests packaged for easy analysis and administration. A tester first selects a number of test sets from the battery. The test sets might, for example, be selected to gain a baseline assessment of the neuropsychological functioning of an individual or to allow the state-specific assessment of neuropsychological function. For example, the ANAM can be used to assess mental function after a concussion, after a period of sleep deprivation, or after exposure to pharmacological agents. ANAM test sets can also be selected to obtain a clinical assessment of medical disorders such as Parkinson's disease or Alzheimer's disease.

The high incidence of IEDs on the contemporary battlefield has heightened the interest of the Army's mission commanders in ANAM. A number of commanders have instituted predeployment baseline testing for all their troops with a subset of ANAM well-suited for identifying neural function changes caused by traumatic brain injury. Such testing allows for comparing an individual's neurological functioning before and after exposures that increase the risk of traumatic brain injury, such as the blast from an IED. The ANAM support center estimates that over 50,000 troops were screened before deployment in 2008.

Summary: Status of Soldier Selection and Assessment and the Potential for Neuroscience-Based Improvements

The ASVAB, which the Army currently uses to assess the fitness of most candidate recruits for the Army and their suitability for specific assignments, does not assess personality traits or neuropsychological traits. Even in the case of high-value assignments for which the Army must make a substantial investment in training, trainees are free to repeat the training sequence until they can meet minimum performance requirements, and the Army knows little about a candidate's neuropsychological traits before the training begins.

To address limitations in the ASVAB, the Army has developed a complementary test that attempts to assess a few basic behavioral traits considered valuable for completing a tour of duty, such as conscientiousness, stress tolerance, and openness to new experiences. For a number of high-value specialties, the Army uses screening tests, but the specificity and sensitivity of these tests have not been evaluated as predictive of performance in the specialties for which they are used. Even when a screening test that incorporates conventional behavior-based factors—for example, the SIFT for helicopter pilot training—has been developed to replace an older test with much less predictive power, the Army has not incorporated the improved test in its recruitment process.

The problem is that even improved tests like the SIFT do not measure the neuropsychological traits that neuroscience research has been identifying. Based on the progress seen when neuropsychological indicators are applied in vocational settings, adding fairly simple neuropsychological testing to the current mix of soldier assessment techniques appears likely to improve soldier selection immensely, particularly for high-value assignments where the training investment is high and individual performance is critical to the accomplishment of a unit's mission.

The ANAM, which is currently deployed and being administered to assess neural function, seems a natural starting point for expanding the Army's selection of testing instruments. As more and more soldiers are screened with ANAM before being deployed in combat theaters, the Army is building a large experience base that, if captured in an analytically useful medium, could provide essential feedback for improving recruit selection processes. If the Army can identify soldiers who excel in their areas of specialization, it could use their ANAM test data to identify characteristics common to high performers but less likely to be found in lower-performing individuals in that area of specialization. The ANAM data could thus be used to develop a set of tools for relating neuropsychological assessments to mission performance.

MONITORING AND PREDICTING SOCIAL AND GROUP INTERACTIONS

The Scope of Social Neuroscience

Most of a soldier's actions involve other people, including fellow soldiers, commanders who are giving orders, the enemy, and noncombatants. Recent findings in social neuroscience can inform and improve Army training, tactics, and leadership for dealing with these social interactions.

Social neuroscience, also called social cognitive neuroscience, investigates the neurophysiological basis for social behaviors. Topics examined are those traditionally studied in social psychology, including attraction to others and attachment, altruism, speech recognition, affiliation, attitudes toward other individuals and groups, empathy, identification of others, kin recognition, cooperation and competition, self-regulation, sexuality, communication, dominance, persuasion, obedience, morality, contagion, nurturance, violence, and person memory. Social neuroscience research integrates information about a person's physiological state, social context, experience, and cognition to understand his or her social behaviors. Studies in social neuroscience draw heavily on findings from affective neuroscience, as many studies have shown that social behaviors have a strong affective (emotional) component (Ochsner and Lieberman, 2001). What social neuroscience adds is explicit attention to brain activity and neurophysiology when studying social behaviors.

The techniques used to measure brain activity directly during experiments involving social tasks include magnetic resonance imaging generally and fMRI in particular, computerized tomography, electroencephalography and magnetoencephalography, positron emission tomography, transcranial magnetic stimulation, and event-related potentials. Other techniques are used as surrogates for brain activity: electrocardiography, electromyography, galvanic skin response, eye tracking, and genotyping. Still other methods used in social neuroscience include drug infusion studies, comparisons with patients who have neurological disorders or focal brain lesions, and comparison with animal models.

Although it is difficult to measure brain activity in a laboratory setting during natural social interactions, techniques that approach a natural interaction have been developed. For example, Montague developed a technique called "hyperscanning" to simultaneously measure brain activity using fMRI while two or more people interact (Montague et al., 2002). More simply, brain activity can be monitored in one person while he or she interacts in real time with one or more individuals whose brain activity is not measured (see, for example, Eisenberger et al., 2003). Many studies use a computer-simulated "person" with whom the study subject interacts without knowing the interaction is with a computer-based simulation. This approach appears to activate the same brain regions associated with actual social behavior (see, for example, McCabe et al., 2001).

Relevance of Social Neuroscience to the Army

There are numerous psychosocial factors that contribute to stress in the armed forces, including unpredictability of danger, concern about resiliency and sustaining performance in combat, inability to control situations, separation from family, recovery from injury, death or injury of comrades, even the anticipation of returning to civilian life. As the neurophysiological conditions that correlate with and perhaps underlie the behavioral and psychological ("mental experience") responses to these and other psychosocial factors are uncovered, it will become easier to understand both their detrimental (stressing) and beneficial (stimulating) consequences. This section highlights several of many applications where research using the methods of social neuroscience is likely to produce results of value to Army training and learning.

Using the methods of social neuroscience, tasks with structured, well-defined parameters of interaction can be used to gauge an individual's commitment to group goals while studying the brain processes associated with individual goals. When the group in question is an Army unit or a surrogate for one, this approach can provide an objective measure of effective training and an individual's likely performance as a team member. By adding factors to the structured interaction—for example, time constraints on decisions, sleep deprivation, environmental stressors, or gunshots and other disturbances typical of combat—group cohesion under stress and the neuropsychological correlates of the observed behaviors can be examined for insights into whether and why cooperation is sustained or fails. Adding neuroscience methods and understanding to established techniques for training and evaluating group interactions and unit performance can improve the diagnostic and predictive value of current practices.

Behavioral measurements obtained through the use of structured interactions, such as measurements based on game theory constructs, can predict how soldiers in a particular unit will interact and respond during combat as well how they will interact and respond within their unit or with other "friendly" groups. Behavioral measurements can be structured to emphasize cooperation, competition, punishment, or a blend of all three. Many structured interactions, or games, have been formulated and studied in the context of behavioral economics for small groups (two to four players). A challenge for the Army will be to extend the results to larger groups in contexts that reflect different points along the full spectrum of warfare. The tasks can be modified to better relate to a field operation, while incorporating the basic structure of a formal game whose behavioral and neurological aspects have previously been established and confirmed.

Social neuroscience research on leadership is a nascent effort that could inform the training of both officers and the soldiers under their command. For example, characteristics of leaders under stress can be studied with the methods of social neuroscience as can the impact of a specific leader's characteristics on the performance of those he or she leads. Behavioral game theory uses insights from human behavior to improve decisions (Camerer, 2003), and, as discussed above, the neural correlates of effective decision strategies are now being mapped. One direct application of this work would be to teach leaders some of the findings from experiments in behavioral game theory in order to improve their awareness of the factors that influence their own decision making behavior as well as that of others. Understanding the behavioral and psychological factors in how choices are made, including the brain mechanisms that support effective choices, can contribute to improving the performance of Army officers and the training regimens for soldiers.

As described in Chapter 8, an important trend in long-term research is the continuing discovery of performance indicators that can be linked to neural state, including biomarkers such as the small biomolecules involved in brain functioning that are relatively easy to monitor. Recent research has shown that when an individual receives an intentional and tangible signal of trust from another individual, the brain releases one of these biomolecules, the neuropeptide oxytocin (OT) (Kosfeld et al., 2005; Zak et al., 2005). OT has been shown to reduce anxiety (Heinrichs et al., 2003). It has been known for some time that OT is associated with bonding to offspring and spouses. Moderate stress also induces the release of OT, but fear and great stress inhibit it. By knowing this brain target, training could be redesigned to induce OT release.

Another aspect of social neuroscience relevant to leadership training concerns the neurophysiological correlates that are being identified for the cognitive constructs by which we interact with other persons as conscious agents. This attribution of a "mind" to others is referred to as the theory of mind (ToM), whereby in order to understand another person and respond to his or her behavior, an individual assumes the other person has a conscious mind directing his or her observed behavior (see Box 3-2).

A large number of studies have allowed brain processes related to ToM attributions to be localized to specific cortical regions (Gallagher and Frith, 2003). Neurophysical monitoring techniques can be used to determine whether these regions of the brain are activated during Army-relevant activities—for example, battle simulation training—to optimize soldier learning and retention. The brain's ToM regions also appear to be active during moral judgments (Young et al., 2007). Neural processes in these regions may affect decisions soldiers make when they apply the Uniform Code of Military Justice to combatants and noncombatants. Thus, the techniques of social neuroscience can be adapted to monitor trainee responses in unaccustomed or problematic situations, to improve training, and to test retention of leadership, group dynamics, and moral judgment skills.

Social neuroscience methods can also enhance Army-relevant research on how group dynamics affect

BOX 3-2
Theory of Mind

ToM, a construct that was introduced by Premack and Woodruff (1978) to characterize the mental ability of higher apes, refers to the ability of individuals to attribute independent mental states to self and others in order to explain and predict behavior (Fletcher et al., 1995). This approach has been particularly important to characterize cognitive development in children (Frith and Frith, 2003) and dysfunction of cognitive development in autism and autism spectrum disorders (Happé et al., 1996) as well as in psychosis (Doody et al., 1998). More recently, several conceptual connections have been made between ToM and other important psychological constructs. For example, it appears that the neural representation of self (Happé, 2003) and, more generally, self-generated beliefs (Leslie et al., 2005) are closely related to the ability of ToM. This proposition is supported by some imaging studies that suggest the importance of the medial prefrontal cortex as part of both self-relevant as well as ToM-related processing (Wicker et al., 2003). Therefore, ToM is an important construct that can be used to examine one's ability to infer mental states, related to self, and beliefs but more important is accessible to experimental modulation using neuroscience approaches.

an individual's ability to process information and make choices based on that information. Most army operations involve soldiers trained to operate as a team. However, in most studies—fMRI, positron emission tomography, and magnetoencephalography—of neural information processing the subjects are isolated from other persons, and the study is not designed to assess differences in performance related to the influence of other team members.

With the increasing importance of network-centric concepts in planning for the future Army, an important issue is the performance of individuals operating as a de facto team that results from being connected by communications links. Social neuroscience methods and insights can help to answer two questions raised by these network-created team situations:

- Do soldiers respond more efficiently to perceived human–human communication than to machine–human communication?
- What is the most efficient size (number of members) needed by a particular team to respond to a specific threat?

Recent findings suggest that a person interacting with another person over a communications link has a different pattern of brain area activation than a person interacting with a computer. In the human–human situation just the dorsolateral PFC is likely to be stimulated, whereas in machine–human interaction the medial PFC also becomes activated (McCabe et al., 2001).

Based on this work, neuroimaging experiments could indicate whether military operators are receiving information from a computer-based system or from another person. In these experiments, which would combine behavioral monitoring and neuroimaging, key variables would be the nature (human or machine) and number of communication channels. One hypothesis is that the responses of team members may vary significantly if, under information overload, they begin to attend in different ways to communications perceived to come from human counterparts and to communications perceived to come from computers, depending on the degree of confidence they have in the two sources. A competing hypothesis is that the team members will in fact operate more efficiently if their sense of social interaction with other humans increases their trust in the information received and encourages them to be concerned about the likelihood of low-probability, high-consequence events.

To explore optimal team size for these de facto teams, experiments could be framed to examine how individuals prune their information load to concentrate on what they perceive to be essential. For example, does a small-world network structure emerge, in which there are highly interconnected groups with sparse connections between groups? (In this type of network structure, each node has a low number of connections compared with the total number of possible connections if every node is fully linked with every other node in the net, but information can pass rapidly between any two nodes in the net.) This question also arises when attempting to maximize the efficiency of road networks, and such research might reveal that in some cases adding a communications channel would actually impede overall information flow.

SUMMARY

Neuroscience techniques (neuroimaging, physiological measures, biochemical assays of brain function) can be used to measure the training status of individual soldiers. Specifically, these techniques can be used to determine (1) functional status, (2) recovery time, or (3) level of training. The sensitivity, specificity, and accuracy of these approaches are unknown and require further research. Thus, a primary short-term goal for the Army should be to conduct this research, which will allow for assessing the training status of soldiers. These efforts should last a relatively short time (<5 years).

There is emerging proof of principle that neuroscience techniques could be useful not just for short-term predictions about outcomes and behaviors but also for predicting the performance, behavior, and potential of individuals over the long term. It is not clear which targets would be most valuable for the Army—for example, the best target population and what outcomes the Army should be most interested in. Close collaboration between neuroscience laboratories and Army leadership would help to develop a common agenda that would benefit both communities. First, the Army would improve its ability to predict performance and, second, the laboratories would obtain further proof of principle of the practicality of their methods. A collaborative effort could be viewed as a long-term initiative with high potential payoff over the next 10 years.

REFERENCES

Baird, A.A., M.K. Colvin, J.D. Van Horn, S. Inati, and M.S. Gazzaniga. 2005. Functional connectivity: Integrating behavioral, diffusion tensor imaging, and functional magnetic resonance imaging data sets. Journal of Cognitive Neuroscience 17(4): 687-693.

Balleine, B.W., N.D. Daw, and J.P. O'Doherty. 2009. Multiple forms of value learning and the function of dopamine. Pp. 367-387 in Neuroeconomics: Decision Making and the Brain. P.W. Glimcher, C.F. Camerer, E. Fehr, and R.A. Poldrack, eds. New York, N.Y.: Academic Press.

Ben-Shachar, M., R.F. Dougherty, and B.A. Wandell. 2007. White matter pathways in reading. Current Opinion in Neurobiology 17(2): 258-270.

Botvinick, M.M., T.S. Braver, D.M. Barch, C.S. Carter, and J.D. Cohen. 2001. Conflict monitoring and cognitive control. Psychological Review 108(3): 624-652.

Brown, J.W., and T.S. Braver. 2007. Risk prediction and aversion by anterior cingulate cortex. Cognitive, Affective, & Behavioral Neuroscience 7(4): 266-277.

Camerer, C.F. 2003. Behavioral Game Theory: Experiments in Strategic Interaction. Princeton, N.J.: Princeton University Press.

Clement, J. 1982. Students' preconceptions in introductory mechanics. American Journal of Physics 50(1): 66-71.

Corrado, G.S., L.P. Sugrue, J.R. Brown, and W.T. Newsome. 2009. The trouble with choice: Studying decision variables in the brain. Pp. 463-480 in Neuroeconomics: Decision Making and the Brain. P.W. Glimcher, C.F. Camerer, E. Fehr, and R.A. Poldrack, eds. New York, N.Y.: Academic Press.

Donovan, C.L., R.E. Mayer, and M.B. Miller. 2007. Individual differences in fractional anisotropy and learning strategy predict variability in brain activity during episodic encoding and retrieval. Society for Neuroscience, 2007 Abstracts. Available at http://www.abstractsonline.com/viewer/viewAbstract.asp?CKey={74138971-902E-45F7-80BF-12CACEC7A65E}&MKey={FF8B70E5-B7F9-4D07-A58A-C1068FDE9D25}&AKey={3A7DC0B9-D787-44AA-BD08-FA7BB2FE9004}&SKey={75963B82-6391-4C70-8E91-F80E73368F20}. Last accessed August 25, 2008.

Doody, G.A., M. Gotz, E.C. Johnstone, C.D. Frith, and D.G. Cunningham Owens. 1998. Theory of mind and psychoses. Psychological Medicine 28(2): 397-405.

Dunbar, K., J. Fugelsang, and C. Stein. 2007. Do naïve theories ever go away? Pp. 411- 450 in Thinking with Data. M. Lovett and P. Shah, eds. Hillsdale, N.J.: Lawrence Erlbaum Associates.

Eisenberger, N.I., M.D. Lieberman, and K.D. Williams. 2003. Does rejection hurt? An fMRI study of social exclusion. Science 302(5643): 290-292.

Fletcher, P.C., F. Happé, U. Frith, S.C. Baker, R.J. Dolan, R.S.J. Frackowiak, and C.D. Frith. 1995. Other minds in the brain: A functional imaging study of "theory of mind" in story comprehension. Cognition 57(2): 109-128.

Frith, U., and C.D. Frith. 2003. Development and neurophysiology of mentalizing. Philosophical Transactions of the Royal Society B: Biological Sciences 358(1431): 459-473.

Gallagher, H.L., and C.D. Frith. 2003. Functional imaging of "theory of mind." Trends in Cognitive Sciences 7(2): 77-83.

Geva, N., R. Driggers, and A. Mintz. 1996. Effects of ambiguity on strategy and choice in foreign policy decision making: An analysis using computerized process tracing. Paper presented at the 30th North American Peace Science Society Meeting, October 25-27. Houston, Tex.: Rice University.

Glimcher, P. In press. Neuroeconomics, general: History. In New Encyclopedia of Neuroscience. Oxford, England: Elsevier.

Graham, H.D., G. Coppin, and M.L. Cummings. 2007. The PRIM: Extracting expert knowledge for aiding in C2 sense & decision making. 12th International Command and Control Research and Technology Symposium, Newport, R.I. Available at http://web.mit.edu/aeroastro/labs/halab/papers/The_PRIM.pdf. Last accessed July 14, 2008.

Happé, F. 2003. Theory of mind and the self. Annals of the New York Academy of Sciences 1001: 134-144.

Happé, F., S. Ehlers, P. Fletcher, U. Frith, M. Johansson, C. Gillberg, R. Dolan, R. Frackowiak, and C. Frith. 1996. 'Theory of mind' in the brain: Evidence from a PET scan study of Asperger syndrome. NeuroReport 8(1): 197-201.

Heinrichs, M., T. Baumgartner, C. Kirschbaum, and U. Ehlert. 2003. Social support and oxytocin interact to suppress cortisol and subjective responses to psychosocial stress. Biological Psychiatry 54(12): 1389-1398.

Hutsler, J.J., W.C. Loftus, and M.S. Gazzaniga. 1998. Individual variation of cortical surface area asymmetries. Cerebral Cortex 8(1): 11-17.

Kording K.P., J.B. Tenenbaum, and R. Shadmehr. 2007. The dynamics of memory as a consequence of optimal adaptation to a changing body. Nature Neuroscience 10(6): 779-786.

Kosfeld, M., M. Heinrichs, P.J. Zak, U. Fischbacher, and E. Fehr. 2005. Oxytocin increases trust in humans. Nature 435(7042): 673-676.

Kosik, K.S. 2003. Beyond phrenology, at last. Nature Reviews Neuroscience 4(3): 234-239.

Leslie, A.M., T.P. German, and P. Polizzi. 2005. Belief-desire reasoning as a process of selection. Cognitive Psychology 50(1): 45-85.

Mayer, R.E. 2003. Learning and Instruction. Upper Saddle River, N.J.: Merrill.

McCabe, K., D. Houser, L. Ryan, V. Smith, and T. Trouard. 2001. A functional imaging study of cooperation in two-person reciprocal exchange. Proceedings of the National Academy of Sciences of the United States of America 98(20): 11832-11835.

McCloskey, M. 1983. Naive theories of motion. Pp. 229-324 in Mental Models. D. Gentner and A.L Stevens, eds. Hillsdale, N.J.: Lawrence Erlbaum Associates.

Mestre, J.P. 1991. Learning and instruction in pre-college physical science. Physics Today 44(9): 56-62.

Miller, M.B., J.D. Van Horn, G.L. Wolford, T.C. Handy, M. Valsangkar-Smyth, S. Inati, S. Grafton, and M.S. Gazzaniga. 2002. Extensive individual differences in brain activations associated with episodic retrieval are reliable over time. Journal of Cognitive Neuroscience 14(8): 1200-1214.

Miller, M.B., and J.D. Van Horn. 2007. Individual variability in brain activations associated with episodic retrieval: A role for large-scale databases. International Journal of Psychophysiology 63(2): 205-213.

Montague P.R., G.S. Berns, J.D. Cohen, S.M. McClure, G. Pagnoni, M. Dhamala, M.C. Wiest, I. Karpov, R.D. King, N. Apple, and R.E. Fisher. 2002. Hyperscanning: Simultaneous fMRI during linked social interactions. Neuroimage 16(4): 1159-1164.

Newell, K.M. 1996. Change in movement and skill: Learning, retention, and transfer. Pp. 393-429 in Dexterity and Its Development. M.L. Latash and M.T. Turvey, eds. Mahwah, N.J.: Lawrence Erlbaum Associates.

Niv, Y., and P.R. Montague. 2009. Theoretical and empirical studies of learning. Pp. 331-351 in Neuroeconomics: Decision Making and the Brain. P.W. Glimcher, C.F. Camerer, E. Fehr, and R.A. Poldrack, eds. New York, N.Y.: Academic Press.

Ochsner, K.N., and M.D. Lieberman. 2001. The emergence of social cognitive neuroscience. American Psychologist 56(9): 717-734.

Paulus, M.P., S.F. Tapert, and M.A. Schuckit. 2005. Neural activation patterns of methamphetamine-dependent subjects during decision making predict relapse. Archives of General Psychiatry 62(7): 761-768.

Poldrack, R.A., F.W. Sabb, K. Foerde, S.M. Tom, R.F. Asarnow, S.Y. Bookheimer, and B.J. Knowlton. 2005. The neural correlates of motor skill automaticity. Journal of Neuroscience 25(22): 5356-5364.

Posner, M.I., and M.K. Rothbart. 2005. Influencing brain networks: Implications for education. Trends in Cognitive Sciences 9(3): 99-103.

Poulton, E.C. 1974. Tracking Skill and Manual Control. New York, N.Y.: Academic Press.

Premack, D., and G. Woodruff. 1978. Chimpanzee problem-solving: A test for comprehension. Science 202(4367): 532-535.

Rajkowska, G., and P.S. Goldman-Rakic. 1995. Cytoarchitectonic definition of prefrontal areas in the normal human cortex: II. Variability in locations of areas 9 and 46 and relationship to the Talairach coordinate system. Cerebral Cortex 5(4): 323-337.

Rakic, P. 2005. Less is more: Progenitor death and cortical size. Nature Neuroscience 8(8): 981-982.

Schmidt, R.A., and T.D. Lee. 2005. Motor Control and Learning: A Behavioral Emphasis. Champaign, Ill.: Human Kinetics.

Squire, L.R., C.E.L. Stark, and R.E. Clark. 2004. The medial temporal lobe. Annual Review of Neuroscience 27: 279-306.

Taylor, M.K., G.E. Larson, A. Miller, L. Mills, E.G. Potterat, J.P. Reis, G.A. Padilla, and R.J. Hoffman. 2007. Predictors of success in basic underwater demolition/SEAL training—Part II: A mixed quantitative and qualitative study. Naval Health Research Center Report No. 07-10. Springfield, Va.: National Technical Information Service.

Taylor, M.K., A. Miller, L. Mills, E.G. Potterat, J.P. Reis, G.A. Padilla, and R.J. Hoffman. 2006. Predictors of success in basic underwater demolition/SEAL training—Part I: What we know and where do we go from here? Naval Health Research Center Report No. 06-37. Springfield, Va.: National Technical Information Service.

Tisserand, D.J., M.P.J. van Boxtel, J.C. Pruessner, P. Hofman, A.C. Evans, and J. Jolles. 2004. A voxel-based morphometric study to determine individual differences in gray matter density associated with age and cognitive change over time. Cerebral Cortex 14(9): 966-973.

Trommershäuser, J., L.T. Maloney, and M.S. Landy. 2009. The expected utility of movement. Pp. 95-111 in Neuroeconomics: Decision Making and the Brain. P.W. Glimcher, C.F. Camerer, E. Fehr, and R.A. Poldrack, eds. New York, N.Y.: Academic Press.

Tulving, E. 2002. Episodic memory: From mind to brain. Annual Review of Psychology 53: 1-25.

Wicker, B., P. Ruby, J.-P. Royet, and P. Fonlupt. 2003. A relation between rest and the self in the brain? Brain Research Reviews 43(2): 224-230.

Yin, H.H., and B.J. Knowlton. 2006. The role of basal ganglia in habit formation. Nature Reviews: Neuroscience 7(6): 464-476.

Young, L., F. Cushman, M. Hauser, and R. Saxe. 2007. The neural basis of the interaction between theory of mind and moral judgment. Proceedings of the National Academy of Sciences of the United States of America 104(20): 8235-8240.

Zak, P.J. 2004. Neuroeconomics. Philosophical Transactions of the Royal Society B: Biological Sciences 359(1451): 1737-1748.

Zak, P.J., R. Kurzban, and W.T. Matzner. 2005. Oxytocin is associated with human trustworthiness. Hormones and Behavior 48(5): 522-527.

4

Optimizing Decision Making

Extensive work in the behavioral sciences and neuroscience over the past decade has highlighted two realities that confront any organization that relies on decisions made by its members. First, decision making by humans is often suboptimal in ways that can be reliably predicted—and sometimes remediated with training (e.g., Kahneman and Tversky, 1979; Ariely, 2008). If a decision maker is explicitly told to make decisions that "maximize financial gain on average" or to "minimize human casualties on average" and fails to do so in a systematic way, social scientists consider the outcome to be a result of suboptimal or inefficient decisions. For example, humans make errors in the ways that they estimate the probabilities of events, react to heavy losses, and deal with ambiguous situations (Kahneman and Tversky, 2000). Suboptimal decision making has been identified in financial circles where billions of dollars are at stake, and it undoubtedly occurs in the Army as well.

Second, individuals differ predictably in their decision making (see, for example, Wu and Gonzales, 1998). These differences can be highly idiosyncratic and are likely tied to stable, long-term traits of personality, physiology, and neurology (see, for example, Weber et al., 2002). This variability in decision making can make an individual better suited for one particular task and less well suited for another.

At present, the Army neither seeks to measure, in a formal way, deviations from optimal or efficient decision making nor attempts to characterize the stable individual differences that shape a particular soldier's decision making. This chapter explores ways in which current and emerging insights from neuroscience, together with neural monitoring tools such as neuroimaging, can help the Army to optimize the decision making of its soldiers and officers in both these senses: first, by identifying and providing countermeasures to suboptimal decision making and, second, by identifying and making optimal use of individual variability in decision-making traits.

THE SOURCES OF SUBOPTIMAL DECISION MAKING

Consider a basketball coach who must transport his team to a distant city and must choose between flying and driving. If he chooses to drive because he believes he is minimizing the risk to his players, he has made a suboptimal decision. Because flying is statistically safer than driving, he is exposing his players to more risk by driving. Indeed, purely from the perspective of rational risk management, even if his team were involved in an aviation accident, he would have made the right decision if he had decided to fly.

The Army could do much more than it currently does to study and correct suboptimal decision making using the full range of neuroscience, including the behavioral and systemic levels. Leader training and assessment has not kept pace with advances in the science. Although decision making is critical to the conduct of warfare, no research unit within the Army focuses specifically on assessing or improving decision making by officers or enlisted soldiers at a behavioral or neurophysiological level. Even high-profile projects such as the Augmented Cognition project (a previous Defense Advanced Research Projects Agency (DARPA) project discussed in Chapter 6) have focused on how to present information rather than on ways to improve decision making per se. Because we now know that human decision making is predictably suboptimal in many situations, the Army must decide whether to optimize the force along this human dimension or ignore an emerging opportunity. Two classes of suboptimal decision making for which there is abundant recent behavioral literature will illustrate how much is at stake in choosing whether to seize on this new understanding or ignore it.

Errors in Assessing Relative Risk

All decision making involves risk, and individuals have different tolerances for risk. As the eighteenth-century mathematician Daniel Bernoulli famously observed, a poor man with a lottery ticket that has a 50 percent chance of winning \$20,000 might very rationally trade that ticket for \$9,000. On average, the lottery ticket he holds is worth \$10,000, but he might choose not to take the risk. Of course how little he is willing to accept for the ticket reflects just how risk averse he is. Someone who is very risk averse might settle for a sure gain of \$1,000 in return for the ticket; someone who is tolerant of risk might refuse even \$9,900. The important point is that there is nothing suboptimal about any of these decisions for Bernoulli's poor man. These different decisions reflect different tolerances for risk that might affect the suitability of an individual for a specific job, but scientists (unlike mission planners) must be silent on notions of optimality with regard to risk tolerance. Of course it may be that individuals who are more risk averse are better suited for peacekeeping missions and individuals who are more risk tolerant are better suited for frontline combat missions, but different tolerances for risk, from the point of view of optimality theory, simply represent different approaches.

The assessment of risk by individuals, however, is an area where the tools of optimal decision making can be brought to bear. Consider a commander ordered specifically to "minimize vehicle losses." Real-life situations seldom provide clear-cut alternatives for which risks and benefits can be estimated with a high degree of certainty. But suppose that a commander has a choice between two plans for committing 500 vehicles to an operation and that his staff has prepared the following best estimates of probable losses:

- For plan 1, there is a 2 percent chance that 50 vehicles will be lost.
- For plan 2, there is a 50 percent chance that 4 vehicles will be lost.

From a decision theoretic point of view, given the goal of minimizing vehicle losses, plan 1 is preferable. If 100 commanders face this decision, all choose plan 1, and the probability estimates are accurate, the losses would be half what they would have been if all had chosen plan 2.[1] Despite the logic behind the probability estimates, behavioral research indicates unambiguously that most people would choose plan 2. For behavioral scientists, this preference reflects a standard feature of decision making, subjective probability distortion (Kahneman and Tversky, 1979). It is now abundantly clear that human decision makers see very small probabilities as much larger than they actually are (for an accessible review, see Plous, 1993). The financial industry is now beginning to take this feature of human behavior into account. Some financial firms now provide their decision makers with training and tools that help them overcome this widespread inefficiency in decision making.

[1]For plan 1, two commanders are likely to lose 100 (2 × 50) vehicles; for plan 2, fifty commanders are likely to lose 200 (50 × 4) vehicles.

Over the past decade, the neurophysiological roots of this kind of behavior have begun to be understood. It seems likely that this enhanced understanding of the mechanisms of decision making will shed new light on the sources of suboptimal decisions. We now know, for example, that structures in the basal ganglia and the prefrontal cortex (PFC) provide valuations for actions, and these valuations appear to be passed to the posterior parietal cortex, among other areas, for decision (see, for example, Glimcher et al., 2005). In fact, recent research has even begun to identify the neural algorithms that lead to some of these inconsistencies (Glimcher et al., 2007).

Loss Aversion in Decision Making

Another factor that leads to seriously suboptimal decision making is the asymmetry with which human decision makers typically weigh risky losses versus risky gains (see Ariely, 2008, and Plous, 1993). Consider a gambler who has \$1,000 in his pocket and is offered a \$100 bet that has a 1 in 20 chance of paying off \$500. With the money in his pocket and the night's action still to come, he reasonably refuses this highly improbable bet. Later on, having lost \$10,000 on credit, he might well accept that same bet. Contemporary studies of decision making suggest that this reflects a change in his risk tolerance. As losses accumulate, decision makers typically become more and more risk tolerant, even to the point of becoming risk seekers. In the financial world, this behavior pattern has led to a number of catastrophic financial events and even to bank failures. Whether the pattern occurs in combat has not been studied formally, but the constancy of human nature and the annals of military history (e.g., Napoleon at Waterloo) suggest that it does occur, with ominous consequences.

Over the past year, the neural basis of this phenomenon has been identified. This finding has broad implications for understanding loss aversion as a source of suboptimal decision making. Before the neurological studies were conducted, the widespread conviction was that loss aversion was a discrete fear response that biased other decision-making mechanisms. The neurophysiological evidence weighs against this view, instead suggesting that a unitary valuation system unrelated to fear drives loss-averse behavior (Tom et al., 2007)

MAKING OPTIMAL USE OF INDIVIDUAL VARIABILITY

The preceding section discussed the evidence that nearly everyone, under some conditions, makes decisions that are less than optimal for achieving the organizational goals set for them. Irrespective of the type of suboptimal decision making, individuals differ in how they make decisions. For example, some individuals are more impulsive than others and some are more tolerant of risk. Such differences do not necessarily mean that risk-tolerant individuals are better decision makers than risk-averse individuals. From an institutional point of view—that is, what is best for accomplishing the Army's mission—a certain decision-making style may be better or less well suited to a given task.

Economic, psychological, and neurophysiological studies conducted over the past several decades have shown that individuals differ in their decision making in predictable ways (see, for example, Wu and Gonzales, 1998) and that these predictable differences can remain stable for long periods of time (see, for example, Weber et al., 2002). In an economic context, the attitudes of an individual to risk are idiosyncratic, but they appear to be relatively stable personality features. Personality features such as "conscientiousness" have been shown to be both measurable and stable across an individual's life span (see, for example, Costa and McCrea, 1985). At a neurophysiological level, the distributions of neurochemicals, receptors, and brain activations differ among individuals, and there is growing evidence that these differences are related to stable features of personality (see, for example, Kable and Glimcher, 2007).

A common theme that emerges from all this work is not that one type of decision maker is inherently better than another in all circumstances but rather that, given the natural range of differences in a population, one set of decision-making characteristics may be better suited to a particular task than another. Consider a commanding officer who must select a lieutenant for command in an area filled with civilians. An individual who is more risk-averse may be better suited for this task than one who is more risk-tolerant. But an action conducted with a large force in a highly uncertain environment may call for a more risk-tolerant individual. Experienced commanding officers certainly know this, and they select officers for missions according to their individual strengths.

Informal assessment occurs routinely throughout the Army, as it does in every organization. The issue is whether adopting more formal techniques based on the results of research in neuroeconomics, neuropsychology, and other neuroscience disciplines can give the Army an advantage in decision making. The Army does not now assess officers for traits such as risk tolerance; nor does it train commanding officers to make the best use of the differential decision-making traits among their subordinates. Battalion-level commanders are not given objective information about the decision-making traits of newly assigned subordinates, nor have those subordinates been trained explicitly in how to adjust their decision making to different tasks. More generally, the Army has not sought to characterize the decision-making characteristics of its individual soldiers and officers through any set of validated metrics even though such information could be useful for (1) determining which tasks an individual appears particularly well suited to perform or (2) identifying training strategies that might better prepare an individual for the decision-making responsibilities of his/her assignment, given that individual's inherent behavioral and neurophysiological characteristics.

Is there a more efficient and reliable way to find the best officer for a particular job? Can the Army routinely provide its commanding officers with a more complete, reliable, and useful assessment of their junior officers at the beginning of a tour of duty? Can neurophysiological monitoring tools identify soldiers who face specific idiosyncratic risks in particular decision environments? Issues such as these do not appear to have been explored by the Army. Based on what the private sector is already doing to remain competitive in a global environment, these and similar questions should no longer be ignored. As the ability to characterize and predict behavior and psychology using the tools of neuroscience grows, these tools may well become critical for the Army.

TOOLS FOR CHARACTERIZING INDIVIDUAL DECISION MAKERS

Both officers and soldiers make decisions that contribute to (or detract from) achieving mission objectives. We know three important things about their decision-making processes: First, human decision making is likely to be suboptimal in the sense discussed above. Second, humans vary in the stable long-term personality traits that suit them for different kinds of decision-making tasks and different kinds of decision-making training. Third, people can be trained to be more efficient decision makers. If the goal of the Army is to optimize decision making in both ways—as a countermeasure to suboptimal decisions and to maximize the value of trait variability—then it must characterize how individuals make decisions. Once individual decision making can be characterized objectively, then training regimes and selection procedures for both kinds of optimization can be applied.

Several tools already exist for characterizing relevant traits of individuals as decision makers. These tools fall under the broad categories of behavioral testing, psychophysiological assessment, neurophysiological monitoring, and genetic analysis. Each tool can provide information about how an individual compares to a larger population of decision makers, individual decision-making strategies, the effects of training on decision making, or combinations of these.

Personality as a Factor in Decision Making

The most well-developed and robust form of assessment available today is personality assessment based on factor analysis. During the 1980s and 1990s a number of tools were developed for classifying personality. Investigators subsequently asked whether the features of personality measured with these tests remained stable across an entire life span. In general, the answer has been yes. Among the well-known and well-validated tools for classifying personality traits quantitatively are the NEO personality inventory[2] and the Minnesota Multiphasic Personality Inventory.

Personality traits as measured by these tools have already been used successfully to predict performance in Army assignments. For example, Lawrence Katz has shown that the results from this kind of assessment can lead to a twofold improvement in selecting helicopter pilot candidates, when success is defined as selecting individuals who will later be assessed by their experienced peers as being good helicopter pilots.[3]

The extension of these quantified personality inventories to characterizing individual decision making is less well established. How individuals differ in stable personality traits that influence their decision making has been the subject of intense study in behavioral economics over the last decade. Individuals differ in their tolerance of risk, and they also differ in how they react to risky situations that involve losses and risky situations that involve gains. Individuals can differ as well in the ways that they deal with ambiguous situations and in how they interpret probabilities. People also differ in impulsivity. Some place great value on immediate gains while others are more patient. We do not yet know how stable these traits are over an individual's life span, but we do know that they have a huge effect on the decisions he or she makes.

Emotional Reactivity in Decision Making

The personality inventories discussed above use subjects' responses to a psychological instrument such as a questionnaire to characterize and quantify personality traits. This is not, however, the only way to characterize aspects of personality relevant to how an individual makes decisions. Over the past decade, galvanic skin response (GSR) has become a valuable tool for quantifying an individual's emotional reactivity. GSR is an inexpensive biometric device that measures changes in skin conductivity (essentially, the rate of sweating). Neuropsychological studies have shown that GSR is a reliable surrogate for a subject's arousal and emotional engagement during a number of behavioral tests (for a review, see Phelps, 2002). Recently, the Army Institute for Creative Technologies implemented a GSR-based system for assessing post-traumatic stress disorder (PTSD). GSR could also be used to monitor an individual's emotional reactivity when making decisions in a range of contexts. For example, a GSR-based monitor could be used during simulation training to select individuals for future high-stress missions such as explosive ordnance disposal.

[2]The acronym NEO derives from the first three of the five axes, or major traits, of personality measured by this tool: neuroticism, extraversion, openness, agreeableness, and conscientiousness. Each axis is further divided into six subtraits.

[3]Lawrence Katz, research psychologist, Army Research Institute, briefing to the committee on April 28, 2008.

Emerging Tools: Genetics, Neurohormones, and Brain Imaging

Genetic markers, neurohormones, and brain imaging are emerging as sources for biomarkers that may prove to be reliable indicators of neural state when individuals make choices—that is, they can signify behavior underlying the emotional or subjective elements during decision making. (See Chapter 2 for the definition of "biomarker.") Genetic markers are particularly relevant for identifying stable traits. Research data suggest that some genetic markers can identify individuals at greater risk of reacting to chemical agents or suffering from PTSD. It is also known that hormonal state—specifically, an individual's hypothalamo-pituitary axis responsivity—influences decision making as well as fitness for duty (Taylor et al., 2006, 2007). The cost of genetic testing for such traits will decrease over the next decade, and the selectivity and specificity of the tests will improve. As this happens, the Army should position itself to take advantage of the new tools.

At present, brain scanning seems too costly relative to the utility of the information that could be gained to be of much use for assessing how people make choices. A single brain scan costs more than $1,000, so this technology is unlikely to be useful for this purpose in the near term. For the far term, its potential value for Army applications related to decision making will depend on the increase in understanding it provides relative to other techniques (e.g., the conventional personality inventories or the improved tools described in Chapter 3) and on its cost. For the time being, the Army should follow the direction of research in this area and should be able to assess the practical value of the science and applications emerging from external parties (academia and the commercial sector).

NEUROSCIENCE-RELATED THEORIES OF DECISION MAKING

No single theory of decision making is accepted throughout the multiple disciplines, subdisciplines, and research paradigms that constitute neuroscience as defined for this report. Even within a single discipline, researchers are working on approaches that partly overlap (agree) and partly compete or conflict with one another. The committee has selected two such approaches from among the many that could be discussed to illustrate how the general theories of decision making are relevant to the Army and which aspects of the ongoing research are worth monitoring to gain insights and practical approaches to improving decision making in military-relevant contexts. The first approach is belief-based decision making and the neurophysiological activation patterns that have been linked with its more theoretical constructs. The second approach is intuitive, or naturalistic, decision making, which thus far has been primarily descriptive (taxonomic) of actual decision behavior and has served as a framework for understanding recognition-primed decisions.

Belief-Based Decision Making

This section reviews one approach to decision making that occurs frequently in ordinary situations and that has profound implications for performance in complex environments. Specifically, belief-based decision making is concerned with selecting one action over another (or others) when knowledge about the possible consequences and outcomes of the alternatives is either uncertain, unclear, or subject to very strong a priori assumptions about how actions "should" result in certain outcomes. The theory of belief-based decision making deserves close scrutiny because many choices that an outsider might call illogical might be better understood when looked at from the perspective of the decider's preexisting belief system. Finally, and most important, belief-based decision making has been associated with brain structures that may be accessible to real-time monitoring—and perhaps even to modification—using the tools of neuroscience.

The Cognitive Psychology of Belief Sets

A belief can be defined as a propositional mental construct that affirms or denies the truth of a state of affairs and is closely linked to basic judgment processes. A large and stable set of beliefs is essential for intelligent behavior, since such a belief set forms the basis for any actions that a person may take to achieve his or her goals. Beliefs are the building blocks we use to build mental models of the state of the world. They are therefore important constructs used to guide decision making.

The neuropsychological approach to characterizing beliefs has focused recently on developing testable process theories, which through neural state monitoring—for example, neuroimaging or event-related potentials—can be related to functions of different systems in the brain. To understand recent neuroscience findings related to belief-based decision making, a few essential constructs need to be explained.

A "theory of mind" (ToM) refers to an individual's everyday ability to attribute independent mental states to him- or herself and to others for the purpose of predicting and explaining behavior (Happé, 2003) (see also Box 3-2). The cognitive mechanisms underlying this ability to attribute a "mind" to oneself or others have been extensively investigated via reasoning tasks that require participants to predict the action of agents based on information about those agents' beliefs and desires. Researchers who have investigated this ability have found systematic reasoning errors (Wertz and German, 2007). For example, we typically infer the beliefs of others from what they tell us. However, behavioral experiments have the advantage that the experimenter does not have to rely on the degree to which an individual is reporting his or her beliefs truthfully. Behavioral experiments show that a false belief engages greater cognitive resources, resulting in longer response times. (In this context, a "false belief" is a presumed state of reality that differs from the experienced state of reality [Apperly et al., 2008] or an expected outcome of an action that differs from the actual outcome [Grèzes et al., 2004].) A false belief may increase the cognitive load because subsequent judgments and decision making need to take it into account as a conflict with the individual's perception of reality (Apperly et al., 2008).

When individuals are asked to make belief-based decisions, they tend to (1) examine less closely those arguments that support their existing beliefs and (2) seek out evidence that confirms rather than contradicts current beliefs. These tendencies often lead to belief-maintaining biases and high irrationality (Evans et al., 1993). Interestingly, repetition of the presentation of a belief has been shown to increase the intensity of that belief (Hertwig et al., 1997), which parallels the observation that individuals prefer stimuli that are frequently presented over those that are rarely presented (Zajonc et al., 1974). Dysfunctional beliefs and associated choices of actions are thought to play a crucial role in various forms of dysfunction, including dysfunctional decision making. Experimentally, researchers have used the Implicit Association Test (Greenwald et al., 1998) to assess belief-based associations, which is thought to be superior to traditional self-reporting measures (De Houwer, 2002). Moreover, the Implicit Association Test can be used to assess the strength of evaluative associations in the domain of beliefs (Gawronski, 2002).

More specific for decision making, people have erroneous beliefs about the laws of chance (Tversky and Kahneman, 1971). Beliefs can have direct effects on the perceived likelihood of certain outcomes. For example, familiarity tends to increase the probability of the outcome associated with an event (Fox and Levav, 2000). Moreover, when judging cause and effect, beliefs can interfere with detecting cause and effect in decision-making situations by favoring more believable causes over those perceived as less believable (Fugelsang and Thompson, 2000). Some investigators have integrated these findings into the context of the ToM literature (Fox and Irwin, 1998), suggesting that probabilities of available options are immediately embedded into a mentalizing context. A person's beliefs about the relative importance of probabilities and values also compete with limitations on his or her ability to make decisions on the basis of these beliefs when processing information in a specific decision-making situation (Slovic and Lichtenstein, 1968). For example, greater deviation from optimal decision making in favor of belief-based decision making can be observed when individuals are making decisions in complex and extreme environments. Finally, beliefs can bias a decision maker to adopt what has been referred to as rule following, which occurs when a rule is applied to a situation to minimize the cognitive effort and provide emotionally satisfying solutions that are good enough but not necessarily the best (Mellers et al., 1998). Thus, it should not be surprising that adding measures of personal belief salience to considerations of the decision-situation characteristics improves the accuracy of predictions concerning decision-making behavior (Elliott et al., 1995).

Neural Correlates of Belief Processing

Based on the preceding brief review of cognitive processes that contribute to belief formation and how belief affects decision making, it should be clear that no single area of the brain is responsible for belief-based decision making, and surprisingly few brain regions have been reported to play the same role repeatedly. Neuroimaging studies reveal a neural system with three components: the medial PFC, temporal poles, and a posterior superior temporal sulcus (Frith and Frith, 2003; Gallagher and Frith, 2003).

The right temporoparietal junction (TPJ) has been implicated not only in processing the attribution of beliefs to other people (ToM) but also in redirecting attention to task-relevant stimuli. Recent studies support the hypothesis that the overlap between ToM and attentional reorienting suggests the need for new accounts of right TPJ function that integrate across these disparate task comparisons (Mitchell, 2008). Others have found that the right TPJ activates differentially as a function of belief and outcome and that the activation is greatest when there could be a negative outcome based on the belief that action of a protagonist would cause harm to others, even though the harm did not occur (Young et al., 2007). A study of brain-damaged individuals reported that, in addition to the frontal cortex, the left TPJ is necessary for reasoning about the beliefs of others (Samson et al., 2004). Thus, the TPJ appears to be crucial for the representation of a mental state associated with belief formation, possibly focused on disambiguating true from false beliefs.

Activation of the medial PFC is often observed with the subject at rest, a time when individuals presumably engage in internally directed attention-processing of self-relevant thoughts (Kelley et al., 2002) and beliefs (Wicker et al., 2003). Some investigators have found that, whereas the medial PFC is recruited for processing belief valence, the TPJ and precuneus are recruited for processing beliefs in moral judgment and mediate both the encoding of beliefs and their integration with outcomes for moral judgment (Young and Saxe, 2008). These findings are consistent with those of others, showing occurrence of greater activation of the anteromedial PFC and rostral anterior cingulate when participants were implicitly making associations consistent with gender and racial biases. Using event-related potentials to investigate the neural substrates of reasoning with false beliefs revealed a specific late negative component that appeared to be located in the middle cingulate cortex and that might be related to conflict or error detection (Wang et al., 2008). Therefore, the medial PFC appears to be important for processing the degree to which beliefs are self-relevant, possibly with emphasis on the correct or incorrect, acceptable or unacceptable, valence of the belief.

The strength with which one holds to a belief is an important regulator of human behavior and emotion. A recent neuroimaging study shows that states of belief, disbelief, and uncertainty differentially activated distinct regions of the prefrontal and parietal cortices, as well as the basal ganglia. The investigators proposed that the final acceptance of a statement as true or its rejection as false relies on more primitive, hedonic processing in the medial PFC and the anterior insula (Harris et al., 2007). Comparing belief-based and belief-neutral reasoning, some researchers have reported activation of the left temporal lobe and the parietal lobe, respectively, and modulations by the lateral PFC in cases of overcoming belief bias and by the ventral medial PFC in cases of succumbing to belief bias (Goel and Dolan, 2003). Examining belief-based decision making in a navigation task under uncertain conditions showed distinct regions of PFC activation, which was consistent with an underlying Bayesian model of decision making that permits efficient, goal-oriented navigation (Yoshida and Ishii, 2006). Thus, the dorsolateral and ventromedial PFC may have important regulatory functions in moderating belief intensity and the degree to which beliefs influence decision making. Additional brain regions may modulate belief-based processing by contributing resources to (1) override beliefs, (2) enforce beliefs, (3) engage in reward-related processing associated with beliefs (that is, how good it feels to be right or how one's own belief-based decision affects others).

This brief review of brain areas that are important for belief-based decision making shows that beliefs are most likely decomposed in the brain into various component processes—for example, self-relevance, valence, simulation, social relatedness, and other yet-to-be-examined components. These components are subsequently processed by various regions of the brain to arrive at preferences that allow selecting an option and engaging in decision making. Understanding these component processes may help to better monitor and potentially modulate belief-based decision making, particularly considering the significant distortions and suboptimal decisions that beliefs can generate.

Potential Neural Targets for Modifying Belief-Based Decision Making

In Army operations, decision making frequently occurs in extreme environments, which place a high demand on an individual's physiological, affective, cognitive, or social processing resources. In other words, extreme environments strongly perturb the body and mind, which in turn initiate complex cognitive and affective response strategies. Belief-based decision making is often a way for the individual to save resources and to select an action that seems to be the best under the circumstances. However, as pointed out above, belief-based decision making frequently results in suboptimal outcomes because the individual fails to adequately take into account chance, value, and other information. Therefore, modulating belief-based decision making provides an opportunity to optimize outcomes.

The main advance that neuroscience has brought to understanding belief-based decision making is the ability to assign component processes to specific brain regions and to potentially target these regions for monitoring and modulation purposes. Although some brain regions are likely to be more difficult to modulate, others, located on the convexity of the brain, are easily accessible to modulations that could in theory be used to alter the degree to which belief-processing occurring in that area influences decisions. For example, the TPJ is easy to find and can be accessed using brain stimulation techniques such as transcranial magnetic stimulation (Fitzgerald et al., 2002). Moreover, both fMRI and event-related potentials may be useful in monitoring the TPJ for reliable biomarkers of belief strength and truth of belief. Finally, using real-time fMRI (Cox et al., 1995), one could monitor this and other brain regions in response to inquiries about various types of beliefs and measure whether interventions were successful in modulating beliefs, similar to very recent experiments aimed at attenuating emotion (Caria et al., 2007) or pain (deCharms, 2007). Although these approaches are in their infancy, they point to unprecedented opportunities to directly and specifically manipulate the brain to alter decision-making processes.

Summary Points for Army Applications

The preceding discussion of belief-based decision making suggests three key opportunities for the Army:

- Belief-based decision making is an important process that occurs in extreme environments and can have profound implications for optimizing behavior and outcomes. Accordingly, understanding such decision making and modulating it is an important neuroscience opportunity for the Army.
- Research on belief-based decision making supports the idea that a few distributed brain systems are responsible for processing beliefs that contribute to selecting appropriate or inappropriate actions. These brain systems can be (1) monitored and (2) potentially modulated or modified in real time during training or preparation exercises to optimize decision making.
- Research directed at understanding and applying belief-based decision making can be categorized as near term (applicable results expected within 5 years), medium term (applicable results expected within 10 years), or far term (applicable results expected in 10-20 years).
    - —In the near term, identify instances when individuals are making inappropriate, belief-based decisions that result in nonoptimal performance.
    - —In the medium term, develop training tools to let individuals know when they are making nonoptimal, belief-based decisions.
    - —In the long term, monitor (and potentially also modulate) in real time the brain systems that contribute to nonoptimal, belief-based decision making. Investigative tools such as transcranial magnetic stimulation and real-time fMRI are not envisioned to be deployed in actual operations but could be used as tools for training those who make high-impact, high-risk decisions (e.g., field commanders).

Intuitive Decision Making and Recognition-Primed Decisions

While advances in military technology do change the tactics, techniques, and procedures of armed conflict, the eighteenth-century military strategist Carl von Clausewitz stated that war is influenced more by human beings than by technology or bureaucracy (von Clausewitz, 1976). Relatively young leaders making decisions at the point of the strategic spear can become the critical factor in whether an engagement is won, lost, or drawn. For the U.S. military, including the Army, the challenge is less to empower these young leaders to make decisions (the Army has already restructured to encourage such behavior) and more to improve the quality of the decisions they make. Military strategists have emphasized that the strength needed to win the next war will be less about kinetic energy than about cognitive power and performance (Scales, 2006).

For many years, the military community believed that the classical decision-making literature (e.g., normative decision theory, statistical decision theory, and decision analysis) appropriately defined rational decision making for military commanders. However, the last two decades have moved toward studying naturalistic decision making (positive, or descriptive, decision theory). Military leaders most often make decisions based on their previous experiences. These experiences can be based on actual combat, combat training (for example, at the Joint Readiness Training Center, the Combined Maneuver Training Center, or the National Training Center), or home station training. Or, leaders may acquire vicarious experience by studying past leaders in military history. Taken together, these experiences arguably build a military commander's base for so-called intuitive decision making.

Unpacking the terms "intuition" and "intuitive" is an arduous task, as different researchers have defined the terms in different ways, contingent on their own specialty. In a recent popular treatment of the topic, Gladwell (2005) referred to the idea of intuition as "blink" for a decision made in an instant. Dijksterhuis and Nordgren (2006) define intuition as a gut feeling based on past experience and believe that intuition plays a critical role in the theory of unconscious thought. Klein (1989) posits that intuition precedes analysis and that battlefield commanders are told they need to trust their own judgment and intuition, but they aren't told how to develop their intuition so that it is trustworthy. Shattuck et al. (2001) found that more-experienced (longer time in service and higher rank) military leaders tended to ask for less information when making decisions than did officers with less experience. Their study of decisions made by military leaders (a process they called "cognitive integration") suggested that experienced leaders show an intuition that leads them to sample fewer sources and to ignore sources they deem unworthy of their attention. Less-experienced officers sampled all of the information sources available and usually as much of each source as was allowed.

A particular focus of work on naturalistic decision making has been a theoretical model called recognition-primed decision making (RPD) (Klein and MacGregor, 1987; Klein, 1989; Hancock, 2007). RPD aims to provide a descriptive model of how people actually make decisions by predicting the kind of strategy they will use (Klein, 1998, 2008). It has been validated in a number of different domains by different research teams, and it has been empirically tested (Klein et al., 1995). In most studies, decision makers have been found to use an RPD strategy for about 90 percent of their difficult decisions. For simpler decisions, the proportion is likely to be higher. Of course, some decisions force a comparison between alternatives (e.g., candidates for a position) that more closely resembles the decision processes advocated in normative decision theory.

Although the RPD model predicts the strategy that will be used, it does not predict the choice the decision maker will make using that strategy. Nevertheless, attempts to predict or model actual choices are likely to benefit from referring to the decision strategy as a starting point. A pure content model that ignores process (the strategy employed) may not prove useful in describing the actual decision process of experienced leaders or in helping less-experienced leaders to make better decisions. One approach to predicting decision choices is to build on the RPD structure and add, for example, pattern repertoire content. Several ongoing attempts to formulate computational versions of the RPD model are employing this approach to predicting the decision as well as the strategy.

REFERENCES

Apperly, I.A., E. Back, D. Samson, and L. France. 2008. The cost of thinking about false beliefs: Evidence from adults' performance on a non-inferential theory of mind task. Cognition 106(3): 1093-1108.

Ariely, D. 2008. Predictably Irrational: The Hidden Forces That Shape Our Decisions. New York, N.Y.: Harper.

Caria, A., R. Veit, R. Sitaram, M. Lotze, N. Weiskopf, W. Grodd, and N. Birbaumer. 2007. Regulation of anterior insular cortex activity using real-time fMRI. NeuroImage 35(3): 1238-1246.

Costa, P.T., and R.R. McCrae. 1985. NEO-PI Professional Manual. Lutz, Fla.: Psychological Assessment Resources, Inc.

Cox, R.W., A. Jesmanowicz, and J.S. Hyde. 1995. Real-time functional magnetic resonance imaging. Magnetic Resonance in Medicine 33(2): 230-236.

De Houwer, J. 2002. The implicit association test as a tool for studying dysfunctional associations in psychopathology: Strengths and limitations. Journal of Behavior Therapy and Experimental Psychiatry 33(2): 115-133.

deCharms, R.C. 2007. Reading and controlling human brain activation using real-time functional magnetic resonance imaging. Trends in Cognitive Sciences 11(11): 473-481.

Dijksterhuis, A., and L.F. Nordgren. 2006. A theory of unconscious thought. Perspectives on Psychological Science 1(2): 95-109.

Elliott, R., D. Jobber, and J. Sharp. 1995. Using the theory of reasoned action to understand organisational behaviour: The role of belief salience. British Journal of Social Psychology 34(2): 161-172.

Evans, J.St.B.T., D.E. Over, and K.I. Manktelow. 1993. Reasoning, decision making and rationality. Cognition 49(1-2): 165-187.

Fitzgerald, P.B., T.L. Brown, and Z.J. Daskalakis. 2002. The application of transcranial magnetic stimulation in psychiatry and neurosciences research. Acta Psychiatrica Scandinavica 105(5): 324-340.

Fox, C.R., and J.R. Irwin. 1998. The role of context in the communication of uncertain beliefs. Basic and Applied Social Psychology 20(1): 57-70.

Fox, C.R., and J. Levav. 2000. Familiarity bias and belief reversal in relative likelihood judgment. Organizational Behavior and Human Decision Processes 82(2): 268-292.

Frith, U., and C.D. Frith. 2003. Development and neurophysiology of mentalizing. Philosophical Transactions of the Royal Society B: Biological Sciences 358(1431): 459-473.

Fugelsang, J.A., and V.A. Thompson. 2000. Strategy selection in causal reasoning: When beliefs and covariation collide. Canadian Journal of Experimental Psychology 54(1): 15-32.

Gallagher, H.L., and C.D. Frith. 2003. Functional imaging of "theory of mind." Trends in Cognitive Sciences 7(2): 77-83.

Gawronski, B. 2002. What does the implicit association test measure? A test of the convergent and discriminant validity of prejudice-related IATs. Experimental Psychology 49(3): 171-180.

Gladwell, M. 2005. Blink: The Power of Thinking Without Thinking. New York, N.Y.: Little, Brown and Co.

Glimcher, P.W., M.C. Dorris, and H.M. Bayer. 2005. Physiological utility theory and the neuroeconomics of choice. Games and Economic Behavior 52(2): 213-256.

Glimcher, P.W., J. Kable, and K. Louie. 2007. Neuroeconomic studies of impulsivity: Now or just as soon as possible? American Economic Review 97(2): 142-147.

Goel, V., and R.J. Dolan. 2003. Explaining modulation of reasoning by belief. Cognition 87(1): B11-B22.

Greenwald, A.G., D.E. McGhee, and J.L.K. Schwartz. 1998. Measuring individual differences in implicit cognition: The implicit association test. Journal of Personality and Social Psychology 74(6): 1464-1480.

Grèzes, J., C.D. Frith, and R.E. Passingham. 2004. Inferring false beliefs from the actions of oneself and others: An fMRI study. Neuroimage 21(2): 744-750.

Hancock, P.A. 2007. Cold calculation or intuitive instinct: Solving the secrets of sustained survival. Pp. 197-206 in The Psychology of Survivor: Overanalyze, Overemote, Overcompensate: Leading Psychologists Take an Unauthorized Look at the Most Elaborate Psychological Experiment Ever Conducted—Survivor! R.J. Gerrig, ed. Dallas, Tex.: BenBella Books.

Happé, F. 2003. Theory of mind and the self. Annals of the New York Academy of Sciences 1001: 134-144.

Harris, S., S.A. Sheth, and M.S. Cohen. 2007. Functional neuroimaging of belief, disbelief, and uncertainty. Annals of Neurology 63(2): 141-147.

Hertwig, R., G. Gigerenzer, and U. Hoffrage. 1997. The reiteration effect in hindsight bias. Psychological Review 104(1): 194-202.

Kable, J.W., and P.W. Glimcher. 2007. The neural correlates of subjective value during intertemporal choice. Nature Neuroscience 10(12): 1625-1633.

Kahneman, D., and A. Tversky. 1979. Prospect theory: An analysis of decision under risk. Econometrica 47(2): 263-291.

Kahneman, D., and A. Tversky, eds. 2000. Choices, Values and Frames. New York, N.Y.: Russell Sage Foundation.

Kelley, W.M., C.N. Macrae, C.L. Wyland, S. Caglar, S. Inati, and T.F. Heatherton. 2002. Finding the self? An event-related fMRI study. Journal of Cognitive Neuroscience 14(5): 785-794.

Klein, G.A. 1989. Recognition-primed decision. Advances in Man-Machine Systems Research 5: 47-92.

Klein, G.A. 1998. Sources of Power: How People Make Decisions. Cambridge, Mass.: MIT Press.

Klein, G. 2008. Naturalistic decision making. Human Factors 50(3): 456-460.

Klein, G.A., and D. MacGregor. 1987. Knowledge Elicitation of Recognition-Primed Decision Making. Yellow Springs, Ohio: Klein Associates, Inc.

Klein, G., S. Wolf, L. Militello, and C. Zsambok. 1995. Characteristics of skilled option generation in chess. Organizational Behavior and Human Decision Processes 62(1): 63-69.

Mellers, B.A., A. Schwartz, and A.D.J. Cooke. 1998. Judgment and decision making. Annual Review of Psychology 49: 447-477.

Mitchell, J.P. 2008. Activity in right temporo-parietal junction is not selective for theory-of-mind. Cerebral Cortex 18(2): 262-271.

Phelps, E.A. 2002. Issues in the cognitive neuroscience of emotion. P. 539 in Cognitive Neuroscience: The Biology of Mind, 2nd ed. M.S. Gazzaniga, R.B. Ivry, and G.R. Mangun, eds. New York, N.Y.: Norton.

Plous, S. 1993. The Psychology of Judgment and Decision Making. New York, N.Y.: McGraw-Hill.

Samson, D., I.A. Apperly, C. Chiavarino, and G.W. Humphreys. 2004. Left temporoparietal junction is necessary for representing someone else's belief. Nature Neuroscience 7(5): 499-500.

Scales, R.H. 2006. Clausewitz and World War IV. Available online at http://www.afji.com/2006/07/1866019/. Last accessed July 23, 2008.

Shattuck, L.G., J.L. Merlo, and J. Graham. 2001. Cognitive integration: Exploring performance differences across varying types of military operations. In Proceedings from the Fifth Annual Federated Laboratory Symposium on Advanced Displays and Interactive Displays. College Park, Md., March 20-22.

Slovic, P., and S. Lichtenstein. 1968. The relative importance of probabilities and payoffs in risk taking. Journal of Experimental Psychology Monographs 78(3 Pt. 2): 1-18.

Taylor, M., A. Miller, L. Mills, E. Potterat, G. Padilla, and R. Hoffman. 2006. Predictors of Success in Basic Underwater Demolition/SEAL Training. Part I: What Do We Know and Where Do We Go from Here? Naval Health Research Center Technical Report No. 06-37. Springfield, Va.: National Technical Information Service.

Taylor, M., G. Larson, A. Miller, L. Mills, E. Potterat, J. Reis, G. Padilla, and R. Hoffman. 2007. Predictors of Success in Basic Underwater Demolition/SEAL Training. Part II: A Mixed Quantitative and Qualitative Study. Health Research Center Technical Report No. 07-10. Springfield, Va.: National Technical Information Service.

Tom, S.M., C.R. Fox, C. Trepel, and R.A. Poldrack. 2007. The neural basis of loss aversion in decision making under risk. Science 315(5811): 515-518.

Tversky, A., and D. Kahneman. 1971. Belief in the law of small numbers. Psychological Bulletin 76(2): 105-110.

von Clausewitz, C. 1976. On War. M.E. Howard and P. Paret, eds. Princeton, N.J.: Princeton University Press.

Wang, Y.W., Y. Liu, Y.X. Gao, J. Chen, W. Zhang, and C.D. Lin. 2008. False belief reasoning in the brain: An ERP study. Science in China Series C: Life Sciences 51(1): 72-79.

Weber, E.U., A.-R. Blais, and N.E. Betz. 2002. A domain-specific risk-attitude scale: Measuring risk perceptions and risk behaviors. Journal of Behavioral Decision Making 15(4): 263-290.

Wertz, A.E., and T.C. German. 2007. Belief–desire reasoning in the explanation of behavior: Do actions speak louder than words? Cognition 105(1): 184-194.

Wicker, B., P. Ruby, J.-P. Royet, and P. Fonlupt. 2003. A relation between rest and the self in the brain? Brain Research Reviews 43(2): 224-230.

Wu, G., and R. Gonzalez. 1998. Common consequence effects in decision making under risk. Journal of Risk and Uncertainty 16(1): 115-139.

Yoshida, W., and S. Ishii. 2006. Resolution of uncertainty in prefrontal cortex. Neuron 50(5): 781-789.

Young, L., F. Cushman, M. Hauser, and R. Saxe. 2007. The neural basis of the interaction between theory of mind and moral judgment. Proceedings of the National Academy of Sciences of the United States of America 104(20): 8235-8240.

Young, L., and R. Saxe. 2008. The neural basis of belief encoding and integration in moral judgment. Neuroimage 40(4): 1912-1920.

Zajonc, R.B., R. Crandall, and R.V. Kail. 1974. Effect of extreme exposure frequencies on different affective ratings of stimuli. Perceptual and Motor Skills 38(2): 667-678.

5

Sustaining Soldier Performance

Recent technological breakthroughs provide quantitative physiological metrics of human attentiveness, performance, and neural functioning to gauge cognitive fitness and degradation to below an individual's baseline performance optimum. Advances in neuroscientific knowledge, coupled with these technology breakthroughs, are suggesting approaches to counteracting a range of stressors that soldiers confront in operational environments. Among these are: metabolic stressors such as dehydration, sleep deprivation, fatigue, and pain; physiological stressors such as injury and trauma; and psychological stressors such as emotional trauma.[1]

This chapter discusses neuroscience advances aimed at sustaining soldier performance before, during, and after battle. Two recurring themes emerge from the committee's presentation here of recent relevant research findings and opportunities for the Army to enhance its current soldier sustainment activities. First, stresses on the soldier affect both mind and body—the brain and the traditional somatic systems and organs. Complex interactions between neurophysiological and conventional physiological responses to a stressor contribute to the degradations of performance that occur during sustained and intense operations. The same brain–body interactivity holds for identifying and using countermeasures to sustain performance despite those stressors.

The second recurring theme covered in this chapter extends a theme found as well in Chapters 3, 4, 6, and 8: the growing importance of individual variability in all of the areas where neuroscience can contribute to Army applications. In this chapter, the insight to take away is that individuals differ markedly in their response to the various stressors described here, just as they differ in their optimal baseline performance, against which degradation due to stressors should be compared. Individuals differ as well in their response to a countermeasure or intervention intended to mitigate a performance deficit, including the extent to which they are helped by a particular intervention or experience undesirable side effects from it.

Finally, parts of the discussion here extend the usual concept of sustainment as it is typically understood in the Army community. That community generally views the time frame for sustainment in terms of the duration of a single extended operation or action—typically up to 96 hours. The performance deficits discussed here often occur during or soon after a single extended operation (the usual time frame for sustainment), but other deficits affect performance over longer time frames of weeks, months, and even years. These longer times are typically associated with Army concepts such as individual soldier resiliency and, at the unit level, recovery and reset. In a sense, then, this chapter covers soldier resiliency and its implications for unit recovery and reset, as well as the sustainment of performance through an entire operation.

MEASURES TO COUNTER PERFORMANCE DEGRADATION

Knowledge of how the body and brain function can serve to counter degradations in performance resulting from physiological and neurophysiolocical stressors. The operational performance of soldiers will benefit from research to develop effective nutritional as well as pharmacological and therapeutic countermeasures to performance degradation from fatigue, sleep deprivation, and other metabolic stressors.

Fatigue

Fatigue is typically described as a failure of performance with time on task; however, its causes are multiple and complex, including muscle overuse, loss of motivation, circadian disruption, poor nutrition, or depression. Both cen-

[1]The committee notes that such psychological stressors are discussed as degradations of performance in Chapter 5. In general, however, the committee considers them (as in Chapters 3 and 4) to be psychological factors, without regard to how they may or may not affect performance on either a group-averaged or individual basis.

tral nervous system (CNS) and peripheral (muscle) factors contribute to the onset of fatigue during prolonged physical exertion. However, the neurophysiological basis of CNS fatigue is not understood. There has been little systematic neuroscience research into the causes of CNS fatigue and potential countermeasures to it. Given the ever-increasing mental demands on today's warfighter, along with emerging evidence showing the potential for nutrition to reduce CNS fatigue during sustained periods of physical and mental exertion, there is much to be gained by applying neuroscience research to improve warfighter performance. At a minimum, the tools should include functional magnetic resonance imaging (fMRI), transcranial magnetic stimulation, more sophisticated behavioral studies in humans, and basic neurochemical and physiological assessments in rodents.[2]

Good evidence is emerging to suggest that CNS fatigue may be caused, in part, by reduced availability of glucose, the main energy source for the brain, and/or an imbalance of neurotransmitters/neuromodulators, including serotonin, dopamine, adenosine, and ammonia. Under some circumstances, an increase of inflammatory cytokines and elevated brain temperature could also play a role (Bautmans et al., 2008; Miller, 2009).

Carbohydrate and/or caffeine feedings during exercise are the most well-established nutritional strategies used to delay both physical and mental fatigue. Less information is available on promising new nutritional strategies such as tyrosine supplementation, which has been shown to benefit mental performance in military-specific situations, and novel food and spice extracts and phytochemicals derived from traditional medicines like quercetin and curcumin. Phytochemicals may work by virtue of their antioxidant and anti-inflammatory activity as well as their ability to provide sustained energy within the brain and muscle. Although claims of enhanced mental performance during long periods of physical or mental stress are made for a range of nutritional supplements, including branched-chain amino acids (BCAAs), ginseng, ginkgo biloba, and choline, there is little scientific support for these claims.

The development of fatigue during sustained periods of physical and mental stress is a complex and poorly understood phenomenon. During prolonged exercise, many factors contribute to the production and onset of fatigue. These factors can operate peripherally—that is, in the muscles—most importantly through depletion of the intramuscular carbohydrate stores and/or through inhibition of adenosine triphosphate (ATP) hydrolysis due to the accumulation of metabolic waste products such as phosphates and hydrogen ions (Davis and Fitts, 1998). They also act centrally in the brain, but the physiological mechanisms of CNS fatigue are just now beginning to be unraveled. Good evidence is emerging that demonstrates how important a role the CNS plays in the processes of fatigue.

Unfortunately, advances in understanding fatigue and its consequences for performance have been held back because physiologists almost exclusively study peripheral factors in fatigue (e.g., those that involve muscle, heart, or blood) in isolation from CNS involvement, whereas psychologists study mental factors (e.g., cognition, mood, vigilance, sleepiness) in isolation from peripheral interactions. Although mind and body are inextricably linked in the onset and consequences of fatigue, there has been very little focus on the neurophysiological basis of the complex interactions between the brain and peripheral factors.

Nowhere is an understanding of the biological mechanisms by which the CNS and peripheral factors in fatigue interact more important than in sustaining today's soldiers. The increased speed, complexity, and lethality of modern warfare make it even more important than in the past to understand how to sustain or enhance physical and cognitive performance. It is also important to maintain mood and motivation as the foundation of both physical and mental performance. Without such an understanding, it will be difficult to move past outdated strategies such as nutrition and exercise training to offset muscle-specific fatigue or caffeine to maintain wakefulness. This section presents a working model of the factors associated with fatigue during sustained periods of physical and mental exertion. It then briefly reviews emerging evidence of the neurobiological basis of fatigue. This new understanding, along with the likelihood that nutrition can play an important role in mitigating CNS fatigue, can provide the foundation for what should become an area of emphasis in neuroscience research and applications relevant to soldier performance.

A working model of the factors involved in fatigue is shown in Figure 5-1. Fatigue results from mental and physical factors that ultimately increase the conscious perception of fatigue and the impairment of mental and physical performance. Physical performance requires not only the capacity of muscle to maintain its force production but also adequate motivation or effort, mental alertness, clarity of thought, decision-making ability, and mood (Davis and Bailey, 1997; Davis, 2000). Neural processes, including higher-level cognitive processing, are important components of the fatigue state, whose symptoms at onset include decreased energy, motivation, arousal, and vigor, as well as increased tiredness, perception of effort, and force sensation. These feelings of fatigue almost always occur before the muscle actually loses the ability to maintain the required force or power output (Hampson et al., 2001). Although highly trained individuals (e.g., superathletes) can persevere for longer in a fatigued state through motivation and willpower, more generally an individual's perception that he or she can persevere and perform as well in a fatigued state as when well rested is not borne out by objective measures of cognitive performance.

[2]Animal studies are necessary to gain understanding of neurobiological mechanisms of fatigue to a degree that is not possible using human trials.

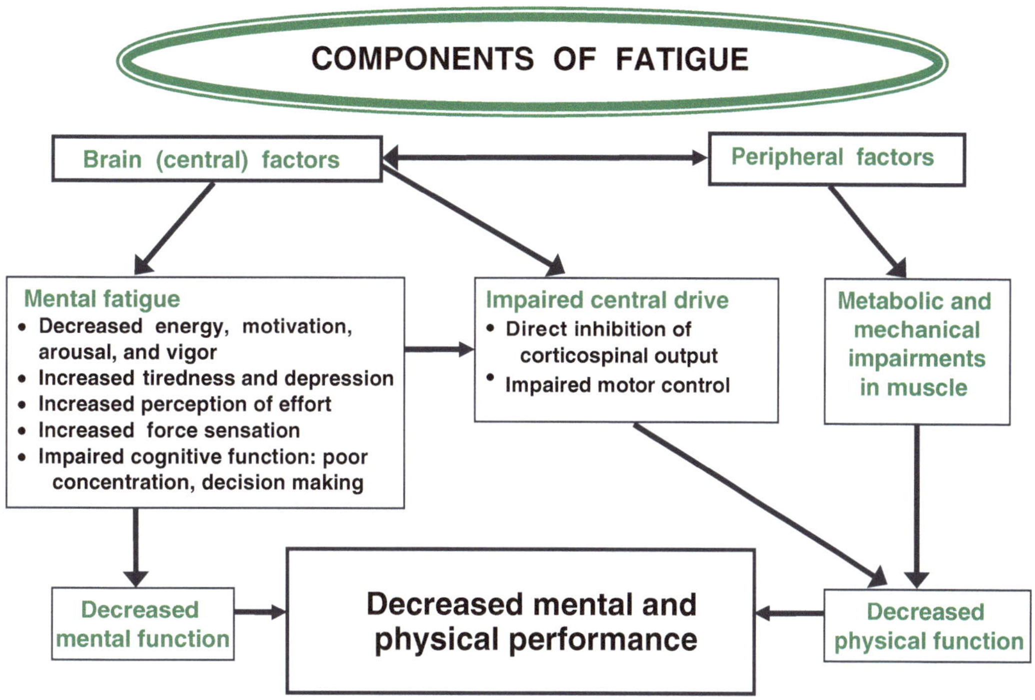

FIGURE 5-1 Schematic diagram illustrating the likely interactions between central and peripheral components of fatigue. SOURCE: Committee-generated.

As fatigue approaches, concentration and decision-making abilities often become impaired. There is also good evidence of direct impairments in central drive to the muscles and of impairments of motor coordination (Gandevia, 2001; Smith et al., 2007; Reis et al., 2008). All these factors are particularly important for today's soldiers, who must have good decision making, vigilance, mood, and motivation for mission success. Maintaining these factors for extended periods of time in harsh environments and at high levels of physiological and psychological stress is essential for soldiers to perform optimally despite these stressors.

New evidence from fMRI and transcranial magnetic imaging have begun to provide a better understanding of the neural correlates of fatigue (Gandevia, 2001; Cook et al., 2007; van Duinen et al., 2007; Reis et al., 2008). However, these studies have not been applied to whole-body exercise during sustained periods of physiological and psychological stress.

Neurophysiological Basis of CNS Fatigue

Very little neuroscience research has focused on the possible biological basis of CNS fatigue. However, recent evidence suggests testable hypotheses about the factors underlying CNS fatigue. These factors include (1) an inadequate supply of energy (glucose), (2) an imbalance among several neurotransmitters/neuromodulators, and (3) elevated brain temperature. One hypothesis is that CNS fatigue occurs during prolonged periods of intense physiological and psychological stress as a result of increased metabolic and oxidative stress in highly active brain regions, decreased availability of glucose, either delivered by the blood or derived from brain glycogen, and increased levels in the brain of 5-hydroxytryptamine (5-HT), coupled with a decrease in brain dopamine and increases in brain adenosine and ammonia. Under certain circumstances, CNS fatigue could also come from increased inflammatory cytokines and elevated brain temperature (Bautmans et al., 2008).

Carbohydrate: Fuel for the Brain The brain is protected by the blood-brain barrier, which selectively allows transport of important nutrients into the brain. However, glucose is essentially the brain's only fuel source. The exercise-induced reduction in blood glucose and in muscle and liver glycogen can contribute greatly to muscle-specific fatigue, but carbohydrate depletion is an even greater problem for the brain, which stores very little glycogen (Evans and Amiel, 1998). Although glucose availability has traditionally been thought to remain relatively constant throughout the brain under most conditions when adequate blood glucose is maintained (Robinson and Rapoport, 1986), recent research suggests that the brain's glucose supply is compartmentalized. As glucose supply and demand change, concentrations change frequently in some areas but not in others (McNay et al., 2001).

Altering brain glucose availability can affect both physical and mental performance. For example, Koslowski et al. (1981) showed that glucose infusion into the carotid artery delayed fatigue in dogs during treadmill exercise. McNay et al. (2000) found that performance of demanding cognitive tasks reduced tissue glucose availability in a specific region of the brain (hippocampus) that is very active during mental functions, even though blood glucose was well maintained. They also showed that intravenous infusion of glucose blocked the decrease in glucose in the hippocampus and improved performance of the cognitive task. Evidence from human studies also shows important benefits of carbohydrate supplementation for mental function, including perceived exertion, vigilance, and mood (Lieberman et al., 2002; Nybo, 2003). Recent elegant studies in humans provide direct evidence for the important role cerebral carbohydrate availability and energy turnover play in physical performance during prolonged exercise (Nybo et al., 2003a, 2003b).

Adenosine, Ammonia, and Dopamine Increased metabolic stress and decreased glucose availability can lead to an increase in brain levels of adenosine and ammonia. Adenosine is a product of the breakdown of ATP, the body's most important energy molecule. Increased levels of adenosine in the brain have been linked to tiredness and sleep (Huston et al., 1996; Porkka-Heiskanen et al., 1997; Porkka-Heiskanen, 1999).

Plasma ammonia concentration can become markedly elevated during prolonged strenuous exercise. Ammonia can easily penetrate the blood-brain barrier and is toxic to the brain. It has been proposed that increased levels of circulating ammonia may play a role in CNS fatigue (Banister and Cameron, 1990; Davis and Bailey, 1997). Direct evidence of exercise-induced increases in brain ammonia and impaired brain function was shown in rats by Guezennec et al. (1998) and, more recently, in humans by Nybo et al. (2005). The latter study found a good association between arterial ammonia concentration, brain uptake of ammonia, cerebral spinal fluid (CSF) ammonia concentration, and perceived exertion during prolonged exercise in human subjects.

A high concentration of brain dopamine is associated with energetic mood, arousal, motivated behaviors, and movement initiation. Control of dopamine is responsible for the effects of many stimulants such as caffeine and ephedrine (Davis, 2000). Dopamine levels in the brain initially increase during endurance exercise and then decrease at the point of fatigue (Bailey et al., 1993). Through various modulatory interactions, increases in serotonin and adenosine play roles in decreasing dopamine levels (Fredholm et al., 1999; Davis, 2000).

Brain Inflammation, Interleukin-1β, and Fatigue Cytokines are an important link between the immune system inflammatory responses and CNS fatigue. During times of inflammation, this cross-talk enables the development of behavioral changes whose goal is to hasten recovery. Interleukin-1β, a potent proinflammatory cytokine and one of the first cytokines upregulated during an inflammatory response, can initiate a host of sickness symptoms, known as sickness behavior, which include poor appetite, changes in sleeping patterns, reduced interest in environment, and, most important, profound fatigue (Dantzer and Kelley, 1989). Fatigue may be initiated by interleukin-1β produced by toxins or disuse damage in the peripherial system, but it is now known that these cytokines are also expressed in the brain (Allan et al., 2001). And it is in the brain that they produce their potent behavioral effects (Kent et al., 1992). Whether peripherally released interleukin-1β enters the brain or transmits its inflammatory signal to the brain via afferent nerves (e.g., the vagus nerve) to initiate fatigue is still under investigation. However, in various inflammatory models, including exercise-induced muscle damage, interleukin-1β-induced behavioral effects, including fatigue, can be blunted by a brain-administered interleuken-1 receptor antagonist, which indicates that the brain is the origin of interleukin-1β-induced fatigue (Bluthe et al., 1997; Carmichael et al., 2006).

Nutritional Measures to Counter the Effects of Fatigue

An interesting and exciting aspect of this understanding of CNS fatigue is the growing scientific evidence that suggests nutrition may be effective in preventing or at least delaying these responses in the brain.

Branched-Chain Amino Acids Newsholme et al. (1987) laid the foundation for one of the first nutritional strategies to delay CNS fatigue. He reasoned that dietary supplementation of the BCAAs valine, leucine, and isoleucine could delay CNS fatigue by offsetting the increase in the ratio of free tryptophan (another amino acid) to BCAA, thereby preventing the typical increase in uptake of tryptophan by the brain during exercise. Limiting tryptophan access would decrease 5-hydoxytryptamine (5-HT) synthesis. Although administration of BCAA in rats prolonged the time to exhaustion (Yamamoto and Newsholme, 2000), positive effects of BCAA ingestion on exercise performance in humans are largely unsubstantiated (van Hall et al., 1995; Strüder et al., 1996; Davis et al., 1999; Davis, 2000; Lieberman, 2003; Cheuvront et al., 2004).

Although physical performance in humans is apparently not influenced by BCAA ingestion, a few studies have reported benefits for cognitive performance and mood. These benefits include improved effort during postexercise psychometric testing (Strüder et al., 1998), improved postexercise Stroop Colour-Word test scores (Hassmèn et al., 1994; Blomstrand et al., 1997), maintained performance in postexercise shape-rotation and figure-identification tasks (Hassmèn et al., 1994), and less perceived exertion during exercise (Blomstrand et al., 1997). However, Cheuvront et al. (2004) reported no significant differences in mood, perceived exertion, or cognitive performance with BCAA supplementation. Similarly, infusion of a saline solution containing BCAA had no effect on perceived exertion during 90-minute treadmill runs (Strüder et al., 1996). The committee believes that there is not enough evidence so far to recommend BCAA supplementation to delay brain fatigue. An important concern here is the well-characterized increase in ammonia production and the brain dysfunction that can occur when large BCAA doses are ingested during exercise (van Hall et al., 1995). In addition, acute tryptophan depletion produces mood and cognitive disturbances in individuals with a history of depression or a heritable variant of the serotonin transporter (Neumeister et al., 2006).

Carbohydrate Supplementation It is well established that carbohydrate supplementation during prolonged, intense exercise delays fatigue. The majority of research to understand this effect of carbohydrate feedings has focused on peripheral mechanisms such as maintaining blood glucose levels and sparing muscle glycogen stores (Coggan and Coyle, 1991). More recently, research has demonstrated the beneficial effects of carbohydrate supplementation on CNS fatigue.

Carbohydrate beverages have been shown to benefit CNS function in both sport and military settings. In the late stages of exercise designed to mimic the demands of soccer and basketball, carbohydrate beverages enhanced performance of gross motor skills, decreased sensitivity to physical force, improved mood (both reported and observed), increased vigor, and reduced perception of fatigue/tiredness. All of these effects were seen in addition to enhanced physical performance, including faster sprint times and higher average jump heights (Welsh et al., 2002; Winnick et al., 2005). Lieberman et al. (2002) also showed a dose-related improvement in vigilance (sustained attention) in a U.S. Army Special Operations unit with carbohydrate supplementation during sustained physical activity designed to mimic a typical combat operation. In addition to improvements in physical performance, volunteers who received carbohydrate supplementation also reported increased vigor and decreased confusion. Even as little as 25 g of glucose can increase cognitive functioning and decrease subjective feelings of mental fatigue (Reay et al., 2006).

The neurobiological mechanisms of the benefits of carbohydrate beverages on CNS function remain speculative. One hypothesis is that carbohydrate beverages provide energy in the form of glucose, along with an increase in dopamine and reductions in brain 5-HT, adenosine, ammonia, and brain temperature. The extra energy supplied by sustaining blood glucose levels presumably lessens the metabolic stress in the brain during exercise, leading to a smaller increase in brain adenosine. The relationship between prolonged exercise, brain adenosine, and tiredness/sleep has recently been demonstrated in rats (Dworak et al., 2007). More glucose can also lessen the increase in brain ammonia, which is toxic to the brain (Nybo et al., 2005).

Greater glucose availability in the blood lessens the concentration of fatty acids in the blood, which in turn blunts the typical increase in free tryptophan available for transport into the brain and presumably attenuates the increase in brain serotonin that is typically associated with tiredness, lethargy, depression, and low arousal (Davis et al., 1992; Bequet et al., 2002; Blomstrand et al., 2005). Lowered serotonin and adenosine combine to combat the drop in brain dopamine as exercise progresses (Davis et al., 2003b) and, along with the reduced levels of brain ammonia, help delay brain fatigue (Davis, 2000; Nybo et al., 2005). Finally, carbohydrates are typically consumed during exercise as sports drinks that provide fluid to offset dehydration and reduce the risk of elevated body temperature—and therefore brain temperature, which is another factor in exercise-induced central fatigue (Nybo, 2008).

Caffeine Caffeine is the most widely consumed nervous system stimulant in the world. It has long been reported to increase wakefulness, subjective feelings of energy, vigilance, mood, and mental and physical performance. All of these effects are due to an increase in dopamine activity and are opposite to those produced by adenosine (Fredholm et al., 1999). Caffeine competes with adenosine for binding to adenosine receptors in the brain, so if caffeine displaces some adenosine molecules, the negative effects of adenosine discussed previously will be reduced (Garrett and Griffiths, 1997). The primary outcome of blocking adenosine receptors with caffeine is a large increase in the release of dopamine in the brain (Fredholm et al., 1999). This hypothesis was confirmed in a recent study by Davis et al. (2003b) of rats during prolonged treadmill running to fatigue.

Caffeine use to improve cognitive functioning during sustained military operations has been extensively researched (NRC, 2001). In a review by Lieberman (2003), it was reported that caffeine improves vigilance in rested and sleep-deprived individuals, improves target detection speed without adversely affecting rifle-firing accuracy, improves decision response (choice reaction) time, improves learning and memory, and improves mood, including perceptions of fatigue and sleepiness. Benefits were seen with as little as 100-200 mg caffeine (~1.4-2.8 mg/kg body weight). Although optimal caffeine doses vary widely among individuals, caffeine can improve cognition and mood in both habitual caffeine users and nonusers (Haskell et al., 2005).

Ephedrine as a Dietary Supplement Ephedrine is another commonly used stimulant that has demonstrated ergogenic properties. Ephedra, which is an extract from the Chinese ma huang plant (an herb in the genus *Ephedra*), contains ephedrine and related alkaloids and is used as a dietary supplement. It has been reported that there is an additive ergogenic effect when ephedrine is used with caffeine, with the performance enhancement being greater than when either stimulant is used alone (Magkos and Kavouras, 2004), and, indeed, most commercial ephedrine-containing products also contain caffeine. A supplement containing 375 mg of caffeine and 75 mg of ephedrine improved run times (Bell and Jacobs, 1999), and Bell et al. (2000) demonstrated that lower doses of caffeine plus ephedrine (4 mg caffeine per kilogram body weight and 0.8 mg ephedrine per kilogram body weight) were just as ergogenic as higher doses but with fewer adverse side effects.

Caffeine, ephedrine, and similar stimulants are associated with adverse side effects, and responses to these drugs are highly individualized. Minor side effects can include insomnia, anxiety, irritability, mild diuresis, headaches, and gastrointestinal distress. More severe side effects can include severe diuresis and dehydration, tachycardia, arrhythmia, hypertension, dependence, seizures, coma, or death. Caution must be taken when using caffeine and/or other stimulants. Sometimes it is difficult to determine that a product contains caffeine or other stimulants because they are not specifically listed as such on the label. For instance, guarana is a popular constituent of many commercial energy products, but it is not widely known that the main active component of guarana is, in fact, caffeine.

Tyrosine Tyrosine is the amino acid precursor of the neurotransmitter dopamine, which has been associated with energetic mood, arousal, motivated behaviors, and movement initiation and control (Davis, 2000). There is promising but preliminary evidence that tyrosine supplementation may have beneficial effects on cognitive functioning and mood, especially under prolonged stress. Chinevere et al. (2002) showed in a study of prolonged and intense cycling that tyrosine lessened perceived exertion. In addition, tyrosine supplementation, when tested in subjects in prolonged stressful environments such as military operations, improved vigilance, choice reaction time, pattern recognition, map and compass reading capabilities, working memory, and mood, including perceptions of fatigue, confusion, and tension (Lieberman, 2003; Magill et al., 2003; Lieberman et al., 2005; Mahoney et al., 2007).

Supplemental dietary tyrosine increases plasma concentrations of the amino acid (Davis and Bailey, 1997), but the mechanism by which this affects brain activity to produce the observed performance benefits is uncertain. For the central catecholamine neurotransmitters—norepinephrine, dopamine, and epinephrine—the rate-limiting step in their synthesis is the activity of the enzyme tyrosine hydroxylase. This enzyme is fully saturated with its tyrosine substrate even under dietary starvation conditions (Cooper et al., 2003). This point argues against a simple mass effect of increased plasma tyrosine on dopamine production. Increased plasma tyrosine may be acting by some other mechanism. For example, it may compete with the amino acid tryptophan for the transporters required to move these amino acids across the blood-brain barrier. A decreased uptake of tryptophan may decrease brain serotonin; the resultant change in the ratio of brain serotonin to dopamine might improve performance when subjects are experiencing CNS fatigue. At present, however, this and other hypotheses for how tyrosine supplementation works remain speculative. Although the evidence is limited, tyrosine supplementation is a promising nutritional intervention that may have significant effects on the CNS, but more studies are needed before specific recommendations can be made.

Herbal Derivatives: Quercetin and Curcumin Herbal remedies for fatigue, both physical and mental, have been used in traditional medicine for centuries. The biologically active components of these remedies are receiving increased attention in the biomedical literature as dissatisfaction grows with the cost and efficacy of conventional Western pharmacological interventions for general ailments. Two of these herbal derivatives that may in time prove to have positive effects on both the mental and physical components of fatigue are quercetin and curcumin.

Quercetin (3,3′,4′,5,7-pentahydroxyflavone) is a flavonoid[3] found in many fruits and vegetables such as apples, onions, red grapes, berries, and black tea. Preliminary evidence suggests that the effect of quercetin on CNS fatigue is like the effect of caffeine as an adenosine antagonist with resulting increase in dopamine activity. Perhaps the most important and exciting aspect of quercetin supplementation is its effects on brain and muscle mitochondrial biogenesis, which would be expected to increase energy availability. Davis et al. (2007a) found that 7 days of quercetin feedings in mice subjected for 3 days to a treadmill run to fatigue (about 140 minutes) increased markers of mitochondrial number and function in muscle and brain compared with untreated controls; they also observed improved running performance in the quercetin-fed mice in both run to fatigue and voluntary wheel-running activities. In human subjects, quercetin feedings increased the maximum aerobic capacity (VO_2 max) and ride time to fatigue on a bicycle as measured by an ergometer (Chen et al., 2008). However, the specific importance of increased brain mitochondria has not been determined in this context.

Another potential application for quercetin in mitigating CNS fatigue involves its powerful antioxidant and anti-inflammatory activity. Quercetin has strong anti-inflammatory properties by virture of its ability to modulate enzymes and transcription factors within pathways essential for inflammatory signaling, including those that are involved in the interleukin-1β signaling cascade (Comalada et al., 2005, 2006; Dias et al., 2005). Quercetin has been shown to reduce the expression of pro-inflammatory cytokines (Huang

[3]The term "flavonoid" refers to a class of secondary plant metabolites often cited for their antioxidant properties. Chemically, a flavonoid is an aromatic compound that has two substituted benzene rings connected by a chain of three carbon atoms and an oxygen bridge.

et al., 2006; Min et al., 2007; Sharma et al., 2007) and other inflammatory mediators such as COX-2 and NF-κB from many types of cells, including astrocytes,[4] following treatment with various inflammatory agents (O'Leary et al., 2004; Martinez-Florez et al., 2005). Quercetin is generally known to have a much wider safety margin when administered for extended periods (Harwood et al., 2007; Lakhanpal and Rai, 2007) than do anti-inflammatory pharmaceuticals, which are sometimes used for off-label medical indications. In short, based on the literature cited above, quercetin may mitigate CNS fatigue during sustained operations and incidentally provide some protection from other stressors such as injury, infection, and toxic exposures.

Understanding the role of inflammation in fatigue is important when examining the possible benefits of another nutritional countermeasure, curcumin. Curcumin (diferuloylmethane) is an anti-inflammatory component of turmeric, an east Asian plant root familiar as a spice in curries but also a traditional herbal medicine used to treat inflammations of arthritis, heartburn and stomach ulcer, and gallstones (NCCAM, 2008). In recent research, curcumin hastened recovery to baseline performance after exercise-induced muscle damage partly because it attenuated the inflammatory response in muscle and presumably also in the brain (Davis et al., 2007b). However, the specific role of curcumin in brain inflammation, interleukin-1β, and fatigue requires further study.

Herbal Products Herbals are a relatively new area of research, and the mechanisms of their actions remain unclear. Ginseng is by far the most studied of the herbal products. Reay et al. (2005) showed that a single 200-mg dose of *Panax ginseng* was associated with improved cognitive performance and lower subjective feelings of mental fatigue on a visual analog scale. In a follow-up study, Reay et al. (2006) tested the possible interaction between *P. ginseng* and a carbohydrate drink. Both the carbohydrate drink and *P. ginseng* showed improvements on some of the cognitive tests and lowered subjective feelings of mental fatigue, but there was no additive effect by *P. ginseng* and glucose on any of the cognitive outcomes measured. It should be noted, however, that ingestion of *P. ginseng* often reduces blood glucose levels, which could be detrimental to cognitive performance.

Because studies of the potential beneficial effects of ginseng on physical exercise performance often involve long-term vs. sporadic/occasional ginseng supplementation, their results are equivocal. Without a reasonable mechanism of action, the evidence that ginseng ingestion can improve exercise performance is not convincing.

[4]Astrocytes are star-shaped, nonneuronal glial cells. Glial cells constitute the essential supporting tissue, or glia (meaning "glue"), of the brain, which is essential for the health and functioning of the neurons (nerve cells).

Guarana (whose active component is caffeine) in commercial supplements is often found in combination with *P. ginseng*. Few studies have assessed the effects of guarana on cognitive and behavioral measures in humans. Haskell et al. (2005) evaluated guarana extract given in doses of 37.5 mg, 75 mg, 150 mg, and 300 mg. Increased secondary memory performance and increased mood (alertness and content) were seen, with the two lower doses showing more beneficial cognitive outcomes than the higher doses. Kennedy et al. (2004) gave subjects single doses of 75 mg guarana extract, 200 mg *P. ginseng*, or a guarana/ginseng mixture (75 mg/200 mg). Guarana supplementation led to improvements in attention tasks, a sentence verification task, and a serial subtraction task. The combination of guarana and ginseng also led to improvements on a speed-of-memory task.

A few studies examined the effects of ginkgo biloba on cognitive effects in healthy young volunteers. Acute administration of ginkgo (120 mg) improved sustained attention and pattern recognition memory but had no effects on mood, planning, mental flexibility, or working memory (Elsabagh et al., 2005). However, 6-week chronic administration of ginkgo biloba (120 mg) had no effect on any of the cognitive or mood variables measured (Elsabagh et al., 2005; Burns et al., 2006). Some studies have shown slight positive cognitive effects of higher doses (360 mg) of ginkgo (Kennedy et al., 2002) and combination ginkgo/ginseng mixtures (Kennedy et al., 2001, 2002).

Choline Supplements Choline serves as the dietary precursor to the neurotransmitter acetylcholine. One study reported that plasma choline levels dropped in marathon participants, and the authors suggested that choline supplementation before or during exercise might improve endurance performance (Conlay et al., 1992). However, there is insufficient evidence to support claims that choline supplementation improves physical or mental performance outcomes. Warber et al. (2000) studied the effects of choline citrate ingestion during a treadmill run to exhaustion, with subjects carrying a load of 34.1 kg. They found that choline supplementation had no effect on time to exhaustion, squat tests, or ratings of perceived exertion. In a similar treadmill test while carrying a load, Deuster et al. (2002) found that choline supplementation (50 mg per kilogram body weight) had no effects on physical or cognitive performance, including tests of reaction time, logical reasoning, vigilance, spatial memory, and working memory.

Summary of CNS Fatigue Countermeasures

Performance during prolonged periods of physiological and mental stress depends not only on the ability to maintain the physical effort required but also the ability to maintain good mental functioning—for instance, to maintain alertness, clarity of thought, decision-making ability, and mood.

Both brain and body contribute to the onset of fatigue during endurance exercise.

Nutritional strategies designed to enhance CNS function are likely to also improve physical performance. CNS fatigue may be caused in part by reduced glucose availability or by an imbalance between serotonin, dopamine, and adenosine, along with an increase in circulating ammonia, inflammatory cytokines, and—sometimes—elevated brain temperature.

The levels of these transmitters as such do not directly change the effectiveness of the neurons that use them. For example, the neuronal circuits that use these transmitters could become unbalanced, but the brain's adaptive systems go to great lengths to maintain equilibrium. What can change in the short term and later drive metabolic compensations is the activity of the neurons that use these transmitters.

Even though our understanding of possible mechanisms for CNS fatigue is severely limited, the evidence has risen to a level that supports development of testable hypotheses and justifies a recommendation to increase systematic neuroscience research into the neurobiological basis of CNS fatigue and possible nutritional countermeasures in both animal and human models. The most promising nutritional strategies involve carbohydrate and caffeine supplementation, which can increase glucose and dopamine while decreasing serotonin and adenosine. Tyrosine supplementation might have some benefits in certain situations, possibly by increasing brain dopamine. Perhaps the most exciting new supplements that deserve more research are quercetin and curcumin. These supplements may work via their antioxidant and anti-inflammatory activity, along with an increase in mitochondrial biogenesis (quercetin only).

To summarize, neuroscience approaches are already being used to examine the effects of commonly used nutritional supplements on brain functioning. These efforts are just the beginning of research on the effects of various candidate nutritional supplements on brain functioning and performance. Although there is ongoing research sponsored or conducted by the Army into the practical effects of nutritional supplements, virtually nothing is known about how or where they affect brain function and which specific aspects of brain functions are improved (or impaired). Therefore, a concerted effort by neuroscience laboratories to support Army-relevant research would help the Army to identify and test potentially useful nutritional supplements. These efforts would probably yield important results within the next 5 years.

Brain Response to Metabolic Stressors

The Army medical community has long sponsored research on the effects of stress, pain, and sleep deprivation on soldier abilities. Neuroscience research has already demonstrated its potential to revolutionize understanding in these areas. For its weight, the brain has an extraordinarily high demand for glucose and oxygen supplied via the general circulation. Although it accounts for only 2-3 percent of total body mass, for an individual in a resting state the brain accounts for 20 percent of blood glucose utilization. The brain exhibits non-insulin-dependent glucose uptake. Glucose is transported from the blood compartment to the astrocytic end feet surrounding the vascular endothelial cells by glucose transport molecules (Dwyer, 2002). Although in the embryonic state the neuronal elements and glial cells contain an abundance of glycogen (a storage form of glucose), the neurons and glia of postnatal humans and adults contain very low levels of glycogen, thus making the brain dependent on vascular sources of energy. Under conditions of glucose deprivation, the brain can use ketone bodies from the breakdown of fats, but a well-fed individual depends almost totally on glucose to fuel the metabolic processes that produce ATP, the intermediary that provides the chemical energy for neuronal processes such as synthesis of macromolecules and cellular repolarization after axonal discharge.

To provide the brain with sufficient metabolic resources, the distance between the small capillaries comprising the vascular bed in the brain approximates 0.1 mm, making the brain one of most highly vascularized organs in the body. Blood glucose levels are regulated by the release of glucose from the liver, where it is produced from glycogen, and by serum insulin levels that are controlled by pancreatic beta cells. Maintaining a functional metabolic state for decision making and other cognitive tasks—including perceptual discrimination, memory recall and new-memory consolidation, rule-based moral judgment, and multiple task performance—that are accomplished through neuronal information processing is thus critically dependent on the supply of energy-producing substrates for ATP synthesis. A variety of neurotransmitters and growth factors affecting neural development and function (for example, thyroxin, galanin [an insulin-like growth factor], estrogen, and corticosteroids) have a measurable effect on glucose uptake and metabolism.

As discussed in the preceding section on countermeasures to fatigue, quercetin is one of the dietary flavonoids under active investigation as a fatigue countermeasure. Quercetin is also of interest as a dietary supplement to sustain brain function and cognition generally. In addition to its antioxidant and anti-inflammatory characteristics, described above, quercetin may have other mechanisms of action (Nieman et al., 2007). It also has a binding affinity to the adenosine A-1 receptor similar to the binding affinity of caffeine (Alexander, 2006). Some of quercetin's reported benefits on brain function and cognition may extend beyond its antioxidant properties, which attenuate the deterioration associated with aging or ethanol intake. By interacting with the adrenergic adenosine A1 receptor, quercetin may improve neural function (Patil et al., 2003; Singh et al., 2003; Naidu et al., 2004).

Various technologies are being studied to enable soldiers to retain complex decision-making capabilities during continuous operations lasting 72 hours. Transcranial electrical stimulation, enhanced contrast iconography, and flashing icons representing threat (red force) on Force XXI Battle Command Brigade and Below video displays are examples of such technologies. In addition, pharmaceutical supplements are in development to sustain complex decision-making capabilities for extended operations.

Sleep Deprivation

Sleep is not the mere absence of wakefulness. It is an active state that is finely regulated and that undoubtedly plays multiple important roles in brain function. While it is accepted that sleep deprivation degrades performance and impairs vigilance, there are other positive functions of sleep that may be degraded when soldiers are deprived of adequate sleep (Van Dongen et al., 2004b).

Current military doctrine is based on operational activities for ground forces continuing for periods up to 72 hours. A persistently high operations tempo places coalition forces in a dominant position with respect to their adversary. However, also associated with 72 hours of sustained operation is sleep deprivation and concomitant reductions in capabilities for decision making and complex cognitive functioning while performing several tasks. The pioneering studies of Gregory Belenky at Walter Reed Army Institute of Research have demonstrated that whereas overlearned sharpshooting capabilities are retained after 40 hours of sleep deprivation, the decision-making ability required for differentiating specific targets from neutral images deteriorates markedly. These and similar observations show that different facets of attention and decision making decay at different rates as sleep deprivation accumulates.

Research over the past two decades indicates that sleep is essential for consolidating memory and promoting synaptic plasticity. Sleep affects the two major types of memory: declarative memory, which involves retention of facts, and procedural memory, which refers to acquisition of complex skills. Declarative memory is encoded, at least initially, in the hippocampus, whereas procedural memory appears to be localized primarily in the frontal cortex (Marshall and Born, 2007). Electrophysiologic studies of place cells (neurons that fire when an animal is in a specific place in a maze) in the hippocampus reveal that the series of place cells activated in the maze during the acquisition of a spatial learning task is reactivated in precisely the same sequence when the animal is in slow-wave sleep but several times more rapidly. Rapid eye movement (REM) sleep, in contrast, appears to benefit the consolidation of procedural memory (Euston et al., 2007). Investigators speculate that sleep gets the brain off-line so that it can engage the diverse cortical circuitry in memory consolidation.

Sleep deprivation severely compromises the ability of humans to respond to stimuli in a timely fashion. The observed deficits have often been attributed to failures of vigilant attention, which many investigators believe is the foundation of the more complex components of cognition. David Dinges at the University of Pennsylvania has exploited the psychomotor vigilance test (PVT)—a high-signal-load, reaction-time test—to characterize the neurocognitive effects of sleep deprivation. As the amount of normal sleep time lost increases, there is an overall slowing of responses and an increase in the propensity to lose focus for brief periods (>0.5 sec) as well as to make errors of commission. During extended periods of sleep deprivation, interactions between the circadian and homeostatic sleep drives contribute to the lapses in performance (Lim and Dinges, 2008).

Chee et al. (2008) used fMRI to study subjects performing the PVT who were either well rested or sleep deprived. They found that lapses occurring with sleep deprivation, compared with lapses after normal sleep, were associated with a reduced ability of frontal and parietal regions to raise activation, dramatically reduced visual sensory cortex activation, and reduced thalamic activation. Notably, correct performance under both conditions elicited comparable levels of frontoparietal activation. Thus, sleep deprivation produces periods of apparently normal neural activation interleaved with periods of depressed cognitive control, visual perceptual function, and arousal. Another finding was that most subjects have poor insight into the degree of their own impairment due to sleep deprivation.

There are substantial traitlike differences among individuals in terms of their vulnerability to neurobehavioral deficits as the amount of lost sleep increases. Three categories have been identified, with Type 1 individuals being highly tolerant of sleep deprivation (cognitive performance near normal on the tests used). Type 2 individuals are those around the mean performance deficit, and Type 3 individuals are substantially more vulnerable to sleep deprivation than the norm (Van Dongen et al., 2004a). These responses to sleep deprivation are highly stable over time for a particular individual, consistent with the response patterns being stable traits with a high genetic component. Identifying reliable biomarkers[5] for these behavioral types is a high priority for the field of sleep research. Screening with the PVT or another suitable biomarker could be useful to the Army in selecting individuals with optimal resistance to sleep deprivation for missions or assignments where sleep time is likely to be below normal for a sustained period.

For soldiers deployed in combat operations, there are profound effects of sleep deprivation on brain glucose utilization (Spiegel et al., 2005). Positron emission tomography studies have demonstrated that sleep deprivation reduces glucose uptake by the brain. Sleep deprivation also affects release of insulin from the pancreas, which affects the levels of circulating blood glucose. The diurnal levels of hormones that respond to food intake (e.g., leptin and ghrelin) are also modified by sleep deprivation. The 28-amino-acid gastric

[5]See Chapter 2 for the committee's definition of "biomarker" and the requirements for a reliable biomarker.

peptide ghrelin also inhibits the pain of inflammatory responses (Sibilia et al., 2006). This pain suppression effect has not yet been correlated with glucose metabolism. The van Cauter group reported that suppression of slow-wave sleep patterns (non-REM sleep) in healthy subjects alters normal glucose homeostasis (Tasali et al., 2008).

Aside from its interest in prolonging wakefulness, the Army would benefit from following the research on the benefits of sleep. It has recently been suggested that short bouts of sleep can facilitate memory consolidation (Lahl et al., 2008). In the future, it may be possible to improve sleep efficiency so that the necessary physiological processes accomplished during sleep could be done much more rapidly or without the loss of consciousness. For example, birds are able to engage in slow-wave sleep in one hemisphere at a time so that they can remain vigilant and even fly. Marine mammals (cetaceans) also experience slow-wave, symmetric sleep in one hemisphere at a time, while the other hemisphere shows wakeful activity (Pillay and Manger, 2004; Lyamin et al., 2008).

The homeostatic facet of sleep-wake regulation is keeping track of changes in "sleep propensity" (or "sleep need"), which increases during wakefulness and decreases during sleep. Increased sleep propensity following extended prior wakefulness (sleep deprivation) is counteracted by both prolonged sleep duration and enhanced non-REM sleep intensity, as measured by electroencephalography. There is compelling and convergent evidence that adenosinergic neurotransmission plays an important role in non-REM sleep homeostasis. Adenosinergic mechanisms modulate individual vulnerability to the detrimental effects of sleep deprivation on neurobehavioral performance. Sleep deprivation increases the levels of extracellular adenosine and of the adenosine A1 receptor in the cholinergic zone of the basal forebrain. Caffeine, an A1 receptor antagonist, prolongs wakefulness and sleep latency by interfering with the rise of sleep propensity during wakefulness, as revealed by the buildup of theta-wave activity over the frontal lobes.

A functional polymorphism in the adenosine-metabolizing enzyme adenosine deaminase contributes to the high interindividual variability in deep slow-wave sleep duration and intensity (Rétey et al., 2005). Additionally, the circadian gene PERIOD 3 has been shown to correlate with differential vulnerability to cognitive deficits resulting from total sleep deprivation (Viola et al., 2007; Groeger et al., 2008). Moreover, caffeine greatly attenuates the electroencephalography markers of non-REM-sleep homeostasis during both sleep and wakefulness.

Whereas the homeostatic process determines sleep needs, the timing of sleep is determined by the circadian process, endogenous cycles of gene expression, and physical activity. The circadian secretion of melatonin from the pineal gland plays an important role in determining sleep onset, and this effect of melatonin has been exploited to manipulate sleep onset in shift workers, for example. A family of genes, known as "clock" genes, has been identified that comprise a "molecular clock." They are expressed in many cell types throughout the body as well as in the neurons of the brain. Notably, clock genes are overexpressed in the cerebral cortex with sleep deprivation, indicating that they also play a role in sleep homeostasis (Tafti and Franken, 2007).

Recent genetic studies reveal a significant role for heritability in sleepiness, usual bedtime, and usual sleep duration. Several genetic loci, including the clock genes, have been identified that mediate this behavior (Gottlieb et al., 2007). These genetic studies point to endogenous, interindividual differences in sleep homeostasis that may need to be identified to optimize selection of soldiers for specific tasks. They also point to potential targets for pharmacologically manipulating sleep in vulnerable individuals.

PHARMACEUTICAL COUNTERMEASURES TO NEUROPHYSIOLOGICAL STRESSORS

Over the past two decades, neuroscience has made remarkable advances in understanding the neural circuitry of memory, drive, mood, and executive function. Furthermore, the neurochemical features that mediate neurotransmission for components of these circuits have largely been characterized. This knowledge has provided the pharmaceutical industry with targets for developing drugs that perturb specific neurotransmitters, with the potential for treating disorders in which these neural systems have been implicated, such as schizophrenia, Alzheimer's disease, severe mood disorders, and critical behaviors affected by specific neuropsychiatric disorders. Prospective neuropharmacological agents that act on wholly novel targets include a nicotinic acetylcholine receptor modulator to improve attention and executive function in attention deficit disorder, N-methyl-D-aspartic acid (NMDA), a receptor-positive modulator to enhance memory consolidation, and a metabotropic glutamate receptor agonist to treat psychosis (Patil et al., 2007).

Over the next 5 to 10 years, it is highly likely that many new classes of drugs will be developed that mitigate symptoms and deviant behaviors associated with neuropsychiatric disorders. Beyond their approved therapeutic indications, these new medications have the potential for sustaining or optimizing the performance of soldiers. In addition, some of them are likely to alleviate the adverse neuropsychological consequences of combat and other extreme stressors, including major depression and stress-related disorders such as post-traumatic stress disorder (PTSD).

As the Army debates using pharmaceuticals that have been approved by the Food and Drug Administration (FDA) for off-label uses such as sustaining or optimizing performance, it needs to consider a number of issues. First, drugs that affect the CNS by acting on a specific neurotransmitter are likely to affect multiple neural circuits, as a particular neurotransmitter is generally used in several functionally distinct circuits, such as dopamine in the striatum modulating

BOX 5-1
Is Salivary Cortisol a Reliable Biomarker?

The levels of cortisol in salivary samples would appear at first glance to serve as an easily applied, dynamic index of the output of the adrenal cortex and therefore of an individual's reaction to environmental or internal stress, broadly defined (Smyth et al., 1998). However, the interpretation of cortisol level for a given person in a particular context is a complex matter. Salivary cortisol measurement kits vary significantly in their ability to obtain paralleling plasma cortisol levels (assumably the gold standard). That issue aside, researchers agree that to get an accurate area-under-the-curve measure of daily cortisol, measurements at 1, 4, 9, and 11 hours after the subject wakes up can provide good coverage.[1]

Single-day assessments are nonetheless very weak approaches to this problem, since cortisol levels are affected by many day-to-day events (Stone et al., 2001). Many factors are thought to be important in cortisol measurement: (1) stable characteristics such as age and gender; (2) state characteristics such as menstrual cycle stage and use of contraceptives and other medications; (3) disease and/or chronic conditions such as liver disease, PTSD, malnutrition or fasting, or lifestyle (e.g., jet lag or shift work); (4) dynamic characteristics such as food intake (e.g., carbohydrates increase cortisol), sleep status (e.g., assess sleep quality and quantity on night prior to cortisol measurement), exercise (e.g., level and timing), and wake-up time; and (5) psychological characteristics such as positive and negative affect, passivity, or coping.

In short, before salivary cortisol can be used as a reliable biomarker (in the sense defined in Chapter 2), a standard method for assessing individual cortisol baseline must be validated. As well, the difference between the individual's baseline cortisol reading and the reading at another time must be validated as a sensitive and specific marker of the biological condition or outcome it is intended to measure.

---

[1]See Web site of the MacArthur Network on Socioeconomic Status and Health at http://www.macses.ucsf.edu/Research/Allostatic/notebook/salivarycort.html. Accessed December 1, 2008.

movement and in the accumbens mediating reward. Thus, it is essential that specificity of action be demonstrated by the development of tools to measure stress on other baseline states desired. Box 5-1 illustrates the challenge of coming up with a tool to measure stress.

Second, one must be concerned about unforeseen or delayed side effects, particularly when medical indications may be present for which the drug has not been formally approved. Cost-benefit analyses must be undertaken using tools to measure baseline states and which aspects of performance are being enhanced and to obtain clinical measures of overall effects, detrimental or positive. It is possible that some components of performance might be degraded even as others are improved. Used in this context, "benefit" refers to enhanced performance—for example, superior ability to withstand sleep deprivation, faster response times, and the overall improvement in carrying out the military mission. It may include a greater likelihood of survival. "Cost" refers not only to dangerous immediate side effects, but also to long-term side effects or even the potential for the enhanced abilities to lead to unacceptably risky behavior or other poor decisions.

Currently, there are a few examples of the use of FDA-approved drugs to sustain behavior or prevent degradation of performance. Modafinil is prescribed to pilots in the Air Force who are tasked to fly prolonged missions. Sertraline hydrochloride (Zoloft, Lustral) is often prescribed to troops who have sustained repeated combat exposure to reduce the consequences of persistent stress and the risk of depression. Nonetheless, the committee has significant concerns about the potentially inappropriate use of performance-enhancing drugs by the military, particularly with respect to whether the benefits outweigh the risks. Still, it may be worthwhile to continue research into the use of neuropharmacological agents to mitigate degraded performance in unique military circumstances when the benefits of the agents outweigh the risks.

To succeed in the area of neuropharmacological countermeasures to performance deficits due to stressors in operational environments, the Army needs to leverage its relationships with entities whose missions are focused on, or at least involve, developing new drugs—these entities include the pharmaceutical industry, the National Institutes of Health, and the university biomedical and pharmacological research communities. The Army should aim to build on the clinical findings of these entities to determine whether a therapeutic, preventive, or optimizing effectiveness purpose of an agent has been established for conditions relevant to Army operations and whether proper administration of a proven-effective agent is both technically feasible and advisable. The Army should use the full range of neuroscience methods to determine the mechanisms of action for a pharmaceutical's proposed use beyond its approved medical indications and to ensure the specificity and selectivity of the proposed intervention. Finally, pharmacogenetics has revealed substantial interindividual variations in drug responses—for example, to

BOX 5-2
Pharmacoimaging with fMRI to Predict Drug Effects

Although there are many gaps in knowledge and challenges in technology that must be bridged before brain imaging can be successfully and routinely applied to monitor soldiers' performance in the field, current imaging techniques can be applied to laboratory-based research directed at answering key questions about drug efficacy. Several recent results support the use of fMRI as a tool for predicting drug effects (Paulus and Stein, 2007; Phan et al., 2008). To function as a test with predictive validity for new treatment agents, a pharmaco-fMRI technique must meet stringent requirements. The insights gained can, however, be readily applied to predicting performance in extreme conditions or assessing the utility of training interventions.

Four steps need to be successfully implemented for pharmaco-fMRI to give valid and useful results: First, one has to identify a brain area that is important for the target process of interest. This area has to be shown to be functionally altered when an individual's performance changes. Second, one has to identify an experimental paradigm that probes (monitors) this brain area. The experimental paradigm should be sensitive to the behavioral effects of anxiety, show no ceiling or floor effects, and be repeatable with negligible learning effects (i.e., have good test-retest reproducibility). It should be simple and relatively independent of volitional effects, be sensitive to basic pharmacological manipulations, activate areas in the brain that are of relevance for anxiety, and show behavioral effects and/or brain-imaging effects that correlate with ratings of anxiety. Third, one has to determine whether there is a correlation between reduction in performance and the BOLD change in the predicted direction with standard interventions (training, etc.). Fourth, one has to demonstrate that the standard pharmacological intervention affects the brain area in the hypothesized direction. Moreover, this effect should show a dose-response relationship (i.e., larger or more frequent doses of the intervention should have a stronger effect).

neuropsychotropic agents. These findings should constrain how widely such drugs are used for any purpose other than their FDA-approved therapeutic indications.

Box 5-2 describes an example of predicting drug effects. Although these approaches are in their infancy, nevertheless they point toward unprecedented opportunities to selectively and specifically manipulate the brain to alter decision making.

BRAIN INJURY

As noted in Chapter 1, the committee was tasked to focus its study on nonmedical applications in light of the numerous studies of medical neuroscience research and applications. Nevertheless, biomedical and neurophysiological knowledge of combat-related brain injury and stress disorders is a prerequisite for assessing opportunities for mitigating these effects of combat, whether through preventive strategies or by prompt and efficient treatment after a soldier has experienced a potentially injurious event. Accordingly, the section begins with a brief overview of the most salient aspects of current biomedical understanding of brain injuries.

The Iraq war has increased awareness and programmatic emphasis on mitigating, preventing, treating, and protecting against neurological damage. This war has seen a marked increase in the risk for traumatic brain injury (TBI) because of the high proportion of soldiers who have been injured by strong explosions due to improvised explosive devices (IEDs) and who have survived because they received prompt medical care. Blasts from IEDs may cause a unique type of brain damage compared with the more typical penetrating injuries of combat (Stuhmiller, 2008). It is also uncertain whether the number and duration of repeated deployments for the same soldiers have contributed to the prevalence of stress-related disorders in veterans of the Iraq and Afghanistan wars.

Neuroscience research has improved our understanding of the brain's response to stress, the pathophysiology of PTSD, and the consequences of TBI. DOD-wide recognition of the importance of neuroscience research in these areas is evidenced by the establishment of the Defense Centers of Excellence for Psychological Health and Traumatic Brain Injury in 2008.

Stress Disorders, Including PTSD

Several studies and reviews have examined the risk for developing PTSD and allied stress-related disorders, such as panic attacks, emotional dyscontrol, and substance abuse (Hoge et al., 2008; Schneiderman et al., 2008; Smith et al., 2008). A recently released RAND study based on a representative sample of nearly 2,000 individuals deployed for Operation Enduring Freedom in Afghanistan and Operation Iraqi Freedom found that 18.5 percent of all returning service members met criteria for PTSD, depression, or both, whereas 19.5 percent reported experiencing a probable TBI during deployment. About a third of those experiencing a TBI had a concurrent mental disorder (RAND, 2008).

Neuroscience research reveals that a complex interaction involving the brain, the adrenal glands, the peripheral automatic nervous system, and the immune system underlies this kind of mental stress (McEwen, 2007). Whereas the acute fight-or-flight response of the stress axis is protective, persistent activation of the hypothalamic-pituitary-adrenal (HPA) axis can have noxious effects on the brain. McEwen coined the term "allostatic overload" to refer to persistent, excessive stress responses. He found that the powerful interaction between high anxiety and impaired sleep (since sleep deprivation causes increased blood pressure) increased levels of cortisol, insulin, and proinflammatory cytokines (McEwen, 2007). Several experimental paradigms have also shown that sleep deprivation inhibits neurogenesis in the hippocampus, which accounts for concurrent subtle cognitive impairments.

Developmental studies of both animals and humans indicate that high levels of stress during human childhood, such as physical or sexual abuse or neglect, can markedly increase vulnerability to stress in adulthood (Heim and Nemeroff, 2002; Kaffman and Meaney, 2007). Stress early in life results in persistent blunting of the HPA axis, which is mediated by a down-regulation of glucocorticoid receptor expression due to methylation of the DNA in the promoter region of the gene (Meaney et al., 2007). Heritable factors such as allelic variants of the gene for the serotonin transporter, which inactivates serotonin at the synapse, can render individuals more vulnerable to stress early in life and at greater risk for depression in adulthood (Lesch and Gutknecht, 2005; Roy et al., 2007).

Brain neurons express the glucocorticoid receptors, which are responsive to mineralocorticoids as well as to glucocorticoids. The hippocampus, a brain structure critically involved in learning and memory, expresses high levels of these corticoid receptors. Thus, it is not surprising that there is a complex relationship among brain glucocorticoid receptor occupancy, behavior, and cognition. The hippocampus, one of the most malleable structures in the brain, exhibits both functional and structural plasticity, including neurogenesis. Chronic stress or chronic treatment with exogenous glucocorticoids is associated with impairments of hippocampal-dependent memory tasks and with reduction in the volume of the hippocampus (Sapolsky, 2003). Persistent stress and elevated corticosteroids suppress neurogenesis and expression of brain-derived neurotrophic factor in the hippocampus. In this regard, quantitative morphometric studies consistently reveal reduced hippocampal volume in patients suffering from PTSD; the degree of atrophy correlates significantly with the degree of cognitive impairment (Bremner, 2006).

It is not yet clear whether the reduced hippocampal volume is a predisposing factor in the development of PTSD or a consequence of traumatic injury. Many young adults (military or civilian) enter their early twenties with neurological change resulting from automobile accidents, football or other athletic injuries, febrile episodes from infection, or drug taking. An abundance of literature attests to the prevalence of all these factors in the U.S. young adult population. Before one can decide whether an individual's small hippocampus was caused by injury incurred during military service and that the injury led to PTSD or whether the small hippocampus predisposed him or her to PTSD, carefully designed studies of the relevant conditions before and after deployment must be conducted. Resolution of this issue has substantial consequences for care delivery, for the value of preventive measures for susceptible individuals, and for compensation related to military service.

Another brain structure that figures prominently in stress-related mood and anxiety disorders is the amygdala. Research by Yehuda and LeDoux (2007) and by Davis et al. (2003a) has established the neuronal pathways that mediate conditioned fear. Conditioning results when information from the conditioned stimulus (a neutral stimulus such as light or sound) converges with information from the unconditioned stimulus (such as pain or a feared object). Processing of feared experience in the lateral nucleus of the amygdala is a critical step in circuitry through which the release of catecholamines, adrenocorticotropic hormone, and cortisol is regulated. Administration of an α-adrenergic receptor antagonist into the amygdala attenuates the development of conditioned fear, whereas administration of exogenous cortisol exacerbates it (Roozendaal et al., 2006). Although catecholamine metabolites increase in subjects with chronic PTSD, most studies indicate lower levels of corticosteroids and a blunted response to stress (Yehuda and LeDoux, 2007). The central role of the amygdala in conditioned fear in experimental animals has been extended to humans via functional brain imaging studies. Exposure to fear-inducing stimuli leads to functional activation of the amygdala. Furthermore, PTSD patients generally demonstrate increased activation of the amygdala in response to a threatening stimulus or even a neutral stimulus, as compared to untraumatized controls or even to traumatized individuals without PTSD (Rauch et al., 2006).

Extinction of conditioned fear may suggest potential treatments for PTSD. Repeated presentation of the conditioned stimulus to an experimental animal without the unconditioned stimulus results in the gradual extinction of conditioned fear. This active process involves new learning and requires the activation of the NMDA subtype of glutamate receptors in the amygdala. When conditioned fear exists, both animal and human studies point to a loss of inhibitory control over the relevant nuclei of the amygdala by the medial prefrontal cortex. Indeed, prolonged stress alters the circuitry linking the medial prefrontal cortex to the amygdala. Davis et al. (2006) demonstrated that the extinction of conditioned fear in experimental animals through repeated exposure to the conditioned stimulus is significantly facilitated by concurrent treatment with a single dose of D-cycloserine, a partial agonist at the glycine modulatory site on the NMDA receptor, which enhances NMDA receptor responses to glutamate.

Controlled clinical trials indicate that reviewing the traumatic experience in a supportive setting (exposure therapy) can be an effective treatment for chronic PTSD, analogous to exposing an experimental animal to the unconditioned stimulus without the conditioned stimulus. Recent studies demonstrate that, analogous to laboratory results with rats, the administration of a single dose of D-cycloserine robustly and persistently enhances the response to cognitive-behavioral therapy (desensitization) in subjects with acrophobia (Ressler et al., 2004). This robust enhancement persists for at least 3 months and involves memory consolidation. Ongoing studies are examining the use of virtual environments that recall soldiers' experiences in a controlled, graded fashion for desensitizing them, coupled with administration of D-cycloserine (Rizzo et al., 2008).[6] These outcomes, along with neuropsychological research connected with recovery/treatment for traumatic brain injuries, are likely to continue as a source of future neuroscience opportunities.

As noted above, sleep deprivation can reinforce the negative cognitive-emotional features of PTSD. An insidious component of PTSD is that anxiety dreams repeatedly awaken some individuals suffering from it. Increased activity of brain noradrenergic neurons may contribute to the pathophysiology of PTSD as well as to the nighttime sleep disturbances and nightmares that accompany the disorder. Increased noradrenergic activity interferes with normal REM sleep and could interfere with the normal cognitive processing of traumatic events. A recent study examined the effects of prazosin, a centrally active alpha-1-adrenergic antagonist. Compared with placebo in a blinded clinical trial, prazosin increased total sleep time by 90 minutes, on average, increased REM sleep duration, reduced trauma-related nightmares, and significantly improved overall clinical symptoms (Taylor et al., 2008).

Not all individuals exposed to trauma develop PTSD. The type of trauma is important; an interpersonal trauma such as rape or combat appears to be more salient than an accident trauma. Other risk factors include lower intelligence quotient, childhood adversity, avoidant personality, and poor social supports. As noted above, reduced hippocampal volume is robustly associated with PTSD, but it is unclear whether it antedates the traumatic events, thereby being an additional risk factor, or is a consequence of trauma and stress.

The higher concordance of PTSD in identical twins compared to fraternal twins supports the involvement of heritable risk factors. Several putative risk genes for PTSD have been identified. In a small study, a polymorphism in the untranslated region of the dopamine transporter gene was associated with greater risk for PTSD in trauma survivors (Segman et al., 2002). The glucocorticoid receptor genotype was found to affect basal cortisol levels in a subgroup of patients with PTSD (Bachmann et al., 2005). Kilpatrick et al. (2007) found interaction between a polymorphism in the serotonin transporter gene, the severity of the trauma, and the level of emotional support when they studied the development of PTSD after a hurricane. This polymorphism has also been associated with increased risk for depression in the context of stressful life events, including childhood abuse (Kaufman et al., 2004). Binder et al. (2008) recently reported that four single-nucleotide polymorphisms of the *FKBP5* gene interacted with the severity of childhood abuse as a predictor of adult PTSD symptoms. *FKBP5* is part of the mature glucocorticoid receptor heterocomplex, which provides face validity for an association.

Major Depressive Disorder in the Military Context

Aside from the experience of IED events, there are a number of other factors related to military service that predispose individuals to clinical depression. Among them are separation from support networks and family, loss or death of close colleagues, divorce or family instability, economic distress, and untoward responses to medication. U.S. soldiers, who are separated from their families for nominal 15-month deployments, are likely to experience all or most of these factors in some form, independent of having received any TBI.

Stress and disruption of the HPA axis are central to the pathophysiology of major depressive disorder. Stressful life events have been associated with the onset of affective illness. A majority of individuals with an episode of major depressive disorder exhibit dysregulation of the HPA axis with resistance to dexamethasone, a potent glucocorticoid receptor agonist. A major depressive disorder is also highly comorbid in patients suffering from chronic PTSD (Scherrer et al., 2008).

Most animal models of depression that are used to screen for antidepressant efficacy are in fact based on acute or recurrent stress. For example, the Porsolt task is one of the most robust predictors of antidepressant efficacy. The duration of swimming when mice or rats are repeatedly placed in a vat of cool water decreases progressively; effective antidepressants restore prolonged swimming. Chronic administration of corticosterone to rats produces a number of signs and symptoms consistent with major depressive disorder, including increased anxiety, shorter latency on the Porsolt test, and impairments in working memory. Thus, although major depressive disorder can occur spontaneously without evident precipitants, in the military context depression may more often be related to persistent stress and trauma.

Resilience

Resilience refers to the ability to successfully adapt to stressors, thereby maintaining psychological well-being in the face of adversity. Recent research focusing on the psychological and neurophysiological underpinnings of resilience should be of considerable interest to the Army because it

[6]Personal communication between Michael Davis, Robert W. Woodruff Professor, Emory University School of Medicine, and Joseph Coyle, committee member.

identifies strategies to enhance resilience, as well as potential indicators of naturally resilient individuals (Haglund et al., 2007). As noted above, research on experimental animals and in humans unequivocally demonstrates that severe stress in childhood such as abuse or neglect renders the individual much more vulnerable to stress in adulthood (Heim et al., 2000; McCormack et al., 2006). However, the experience of modest, controllable stress in childhood results in greater ability to regulate stress responses in adulthood. This phenomenon, known as "stress inoculation," has been demonstrated in experimental animals and humans (Rutter, 1993). Studies in monkeys indicate that infant monkeys subjected to stress inoculation (e.g., in childhood, separation from the mother for an hour every week) display lower levels of anxiety later in life. They have lower basal cortisol levels and enhanced prefrontal cortex-dependent cognition than controls not subject to the stress inoculation (Parker et al., 2004, 2005).

Small-molecule markers for resilience have also been identified. Dehydroepiandrosterone (DHEA) is an endogenous steroid that elevates mood and counteracts the effects of high cortisol. Soldiers subjected to the stress of survival training, who exhibit superior performance, have a higher ratio of DHEA to cortisol (Morgan et al., 2004). In animal experiments, the neuropeptide Y (NPY) has anxiety-decreasing effects and counteracts the behavioral effects of corticotrophin-releasing hormone. Consistent with these results in animals, serum NPY levels of soldiers subject to the stress of survival training correlated positively with performance, suggesting that NPY may be involved in enhanced stress resilience in humans (Morgan et al., 2000).

Studies have identified several attitudes and behaviors that foster psychological resilience to stress, including optimism, active coping, cognitive flexibility, moral compass, physical exercise, and social support (Haglund et al., 2007). Twin studies indicate that temperamental features such as optimism or neuroticism (tendency toward neurotic responses and behavior) are substantially heritable (Wray et al., 2008). Furthermore, neurophysiological research suggests mechanisms that may explain the protective features of behaviors that foster resilience, such as the finding that physical exercise promotes expression in the brain of brain-derived neurotrophic factor (Cotman and Berchtold, 2002). Similarly, social support appears to modulate the HPA axis (Heinrichs et al., 2003).

Longer-Term Performance Deficits Linked to Traumatic Brain Injury

As noted by Hoge et al. (2008), TBI has been labeled the signature injury of the wars in Iraq and Afghanistan, with 15 percent of soldiers deployed to these theaters reporting blast injuries sufficiently severe to result in loss of consciousness or altered mental status. The risk of comorbid PTSD with CNS symptoms was three to four times greater in these soldiers than in those with no history of blast injury. The presence of PTSD and depression are robust predictors of poor physical health and persistent impairment. Yet the biological basis for this association remains poorly understood. It is simplistic to conclude that the co-occurring PTSD is "a psychological response." Neuroscience should provide some leads about the underlying pathology of blast-induced TBI and opportunities for prevention and treatment. Box 5-3 introduces a new area of neuroscience that may help in following up on such leads. There is debate about whether brain injury as a consequence of blast waves differs from brain injury due to penetrating wounds (Bhattacharjee, 2008). Indeed, it appears that some penetrating brain injuries may reduce the risk for PTSD (Koenigs et al., 2008; Sayer et al., 2008).

Research on depression that occurs after a stroke may be particularly relevant to thinking about TBI. Many clinicians believed that post-stroke depression was a predictable psychological response to disability, although studies indicated that comparable levels of disability from other causes did not result in equally high rates of depression. Furthermore, left anterior lesions pose greater risk for persistent depression than right posterior lesions. Comorbid depression with stroke was a robust predictor of poor outcome, especially death. Both central noradrenergic and serotonergic neuronal systems have figured prominently in the pathophysiology of major depression, as they are the targets of action of effective antidepressants. Robinson and Coyle (1980) demonstrated that because of the peculiar trajectory of these fine, unmyelinated aminergic fibers, which project in an anterior to posterior orientation in the cortex, a stroke lesion in the anterior cortex denervates the rostral cortex of their aminergic innervation. Treating post-stroke depression with antidepressants ameliorates the depression and improves the clinical outcome.

Multiple sclerosis is another disorder associated with a high risk of comorbid depression. The multifocal lesions in the case of multiple sclerosis may have devastating effects on the fine aminergic intracortical axons, as demonstrated by the observation that the severity of depression correlates inversely with the CSF levels of 5-hydroxyindoleacetic acid, a metabolite of serotonin.

Given the evidence that exposure to explosion increases the risk for PTSD in the absence of an acute alteration in mental state, the sport of boxing is germane to the discussion of TBI. Markers of cellular damage are increased in the CSF of boxers with no evidence of concussion. The absence of structural magnetic resonance imaging changes in boxers demonstrates that it may be an insensitive index of damage. For example, only 14 percent of 49 professional boxers subjected to structural magnetic resonance imaging showed abnormalities. In contrast, diffusion tensor imaging revealed robust differences between boxers and matched controls, with reduced diffusion and anisotropy, consistent with disruption of axon terminals. Similar findings have

BOX 5-3
Connectomics and Neural Pathway Degeneration

Connectomics, the study of the brain's neural pathways for information transfer, is an emerging area that addresses fundamental issues in how the brain processes information. The name "connectomics" refers to the concept of considering the entire collection of neural pathways and connections as a whole, analogous to viewing the collection of genes in a human cell nucleus as the genome. An emerging technology known as diffusion tensor imaging is used as an enabling technology for the new field.

An example of how research on connectomics could be relevant to the Army is the problem of accounting for the progression of diffuse axonal injury resulting from a blast trauma. The current understanding is that pressure waves produced by the blast propagate across soft tissue interfaces in the brain, creating a shear force that degrades the junctions between white and gray matter. After the immediate physical effect of the blast—and even when no overt signs or symptoms of damage are observed, as in mild TBI—a degenerative pathology often develops over time. The effects of this neural pathway degeneration eventually lead to symptoms that appear months to years after the injury: short-term memory loss, degraded affect, and depression. In extreme cases, patients suffer from Parkinson's-like tremors like those that boxers develop (Erlanger et al., 1999; Jordan, 2000; Toth et al., 2005). This combination of symptoms, with others, may present as PTSD. If this progressive degeneration occurs because neurons are lost along the neural pathways connecting to the cells or junctions that were damaged directly by the pressure waves from the blast, then connectomics may help to explain how the cell loss spreads from the initial foci of damage to other brain regions.

An example of progress in connectomics with long-term relevance for TBI is a recent technique by which researchers can trace individual neural pathways in the brains of transgenic mouse models. While the brain of the mouse embryo is still developing, multiple copies of modified genes are transferred into the cells that will develop into the brain. These genes produce three proteins that fluoresce in yellow, red, or cyan, producing a palate of nearly 100 colors that are randomly distributed. When the mouse is mature, the brain tissue is excised and its neural pathways can be traced by the color coding (Lichtman et al., 2008). This technique is appropriate only for animal models such as these BrainBow mice (see Figure 5-3-1), which are used to learn about the brain's "wiring diagram" and how it develops. In time, however, this fundamental knowledge should contribute to understanding how the brain normally processes information and what happens when disease or injury progressively degrades the neural pathways.

One hypothesis about the long-term effects of TBI is that the white-matter networks are injured. Experiments with BrainBow mice or similar animal models exposed to IED-like blast effects might in the future allow optical measurement of how the injury progresses. Since the immediate effects of a blast trauma can be observed in the field, experiments with blast effects on these animal models might provide proof-of-concept laboratory evidence for whether and how battlefield treatments and neuronal protection technologies could mitigate the immediate blast damage.

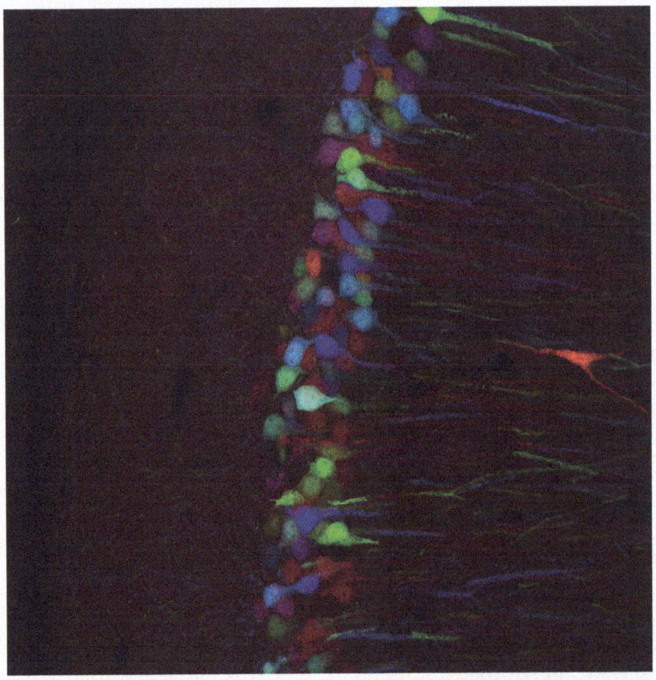

FIGURE 5-3-1 Neuronal pathways in BrainBow mice. Neurons in the hippocampus, a brain area involved in memory, are labeled in different colors, with their neuronal outgoing projections pointing to the left. This is the first time so many different neurons have been separately visualized on such a large scale. SOURCE: Jean Livet, Joshua R. Sanes, and Jeff W. Lichtman, Harvard University (2008).

been obtained in a small study of adolescents with mild TBI who demonstrated an increased anisotrophy and decreased diffusivity at 6 days after an incident.

Little is known about intrinsic risk factors that may affect the outcome of TBI. Apo E4 status robustly predicts neurological deficits in boxers and is linked to poor neurological outcome after TBI from any cause (Jordan et al., 1997; Zhou et al., 2008). Recently, Chan et al. (2008) examined whether polymorphism in the serotonin transporter gene, previously linked to increased risk for depression after psychological trauma, affected the risk for depression and TBI, but they found no association.

Prospective Interventions

While the long-term treatment of PTSD and the consequences of TBI may not be the primary responsibility of the Army, it is in the Army's interest to understand the pathophysiology of these conditions sufficiently to develop effective preventive interventions or acute treatments that mitigate a trauma. Such interventions could include physical training, psychological methods, or pharmaceuticals. Continued research on the identification of risk factors for the development of PTSD could inform interventions that mitigate a risk for PTSD and related stress disorders, thereby lessening the risk of performance deficits and disabilities.

Understanding the neuropathology of TBI should aid in developing better types of body armor to lessen, to the extent possible, the specific types of trauma associated with blast injury. These interventions could hasten recovery after blast exposure and the more rapid return of soldiers to mission. For example, given a known risk factor for PTSD, it may be feasible to reduce the factor for individuals at high risk. Furthermore, it may be possible to develop interventions other than body armor that are suitable for operational use and that would mitigate both the physical and neuropsychological consequences of a trauma.

REFERENCES

Alexander, S.P.H. 2006. Flavonoids as antagonists at A_1 adenosine receptors. Phytotherapy Research 20(11): 1009-1012.

Allan, S.M., D.C. Harrison, S. Read, B. Collins, A.A. Parsons, K. Philpott, and N.J. Rothwell. 2001. Selective increases in cytokine expression in the rat brain in response to striatal injection of α-amino-3-hydroxy-5-methyl-4-isoxazolepropionate and interleukin-1. Molecular Brain Research 93(2): 180-189.

Bachmann, A.W., T.L. Sedgley, R.V. Jackson, J.N. Gibson, R.M. Young, and D.J. Torpy. 2005. Glucocorticoid receptor polymorphisms and post-traumatic stress disorder. Psychoneuroendocrinology 30(3): 297-306.

Bailey, S.P., J.M. Davis, and E.N. Ahlborn. 1993. Neuroendocrine and substrate responses to altered brain 5-HT activity during prolonged exercise to fatigue. Journal of Applied Physiology 74(6): 3006-3012.

Banister, E.W., and B.J.C. Cameron. 1990. Exercise-induced hyperammonemia: Peripheral and central effects. International Journal of Sports Medicine S2: S129-S142.

Bautmans, I., R. Njemini, H. Predom, J.-C. Lemper, and T. Mets. 2008. Muscle endurance in elderly nursing home residents is related to fatigue perception, mobility, and circulating tumor necrosis factor-alpha, interleukin-6, and heat shock protein 70. Journal of the American Geriatrics Society 56(3): 389-396.

Bell, D.G., and I. Jacobs. 1999. Combined caffeine and ephedrine ingestion improves run times of Canadian Forces Warrior Test. Aviation, Space, and Environmental Medicine 70(4): 325-329.

Bell, D.G., I. Jacobs, T.M. McLellan, and J. Zamecnik. 2000. Reducing the dose of combined caffeine and ephedrine preserves the ergogenic effect. Aviation, Space, and Environmental Medicine 71(4): 415-419.

Bequet, F., D. Gomez-Merino, M. Berthelot, and C.Y. Guezennec. 2002. Evidence that brain glucose availability influences exercise-enhanced extracellular 5-HT level in hippocampus: A microdialysis study in exercising rats. Acta Physiologica Scandinavica 176(1): 65-69.

Bhattacharjee, Y. 2008. Shell shock revisited: Solving the puzzle of blast trauma. Science 319(5862): 406-408.

Binder, E.B., R.G. Bradley, W. Liu, M.P. Epstein, T.C. Deveau, K.B. Mercer, Y. Tang, C.F. Gillespie, C.M. Heim, C.B. Nemeroff, A.C. Schwartz, J.F. Cubells, and K.J. Ressler. 2008. Association of *FKBP5* polymorphisms and childhood abuse with risk of posttraumatic stress disorder symptoms in adults. Journal of the American Medical Association 299(11): 1291-1305.

Blomstrand, E., P. Hassmèn, S. Ek, B. Ekblom, and E.A. Newsholme. 1997. Influence of ingesting a solution of branched-chain amino acids on perceived exertion during exercise. Acta Physiologica Scandinavica 159(1): 41-49.

Blomstrand, E., K. Moller, N.H. Secher, and L. Nybo. 2005. Effect of carbohydrate ingestion on brain exchange of amino acids during sustained exercise in human subjects. Acta Physiologica Scandinavica 185(3): 203-209.

Bluthe, R.-M., R. Dantzer, and K.W. Kelley. 1997. Central mediation of the effects of interleukin-1 on social exploration and body weight in mice. Psychoneuroendocrinology 22(1): 1-11.

Bremner, J.D. 2006. Stress and brain atrophy. CNS & Neurological Disorders—Drug Targets 5(5): 503-512.

Burns, N.R., J. Bryan, and T. Nettelbeck. 2006. Ginkgo biloba: No robust effect on cognitive abilities or mood in healthy young or older adults. Human Psychopharmacology: Clinical & Experimental 21(1): 27-37.

Carmichael, M.D., J.M. Davis, E.A. Murphy, A.S. Brown, J.A. Carson, E.P. Mayer, and A. Ghaffar. 2006. Role of brain IL-1β on fatigue after exercise-induced muscle damage. American Journal of Physiology: Regulatory, Integrative, and Comparative Physiology 291(5): R1344-1348.

Chan, F., K.L. Lanctôt, A. Feinstein, N. Herrmann, J. Strauss, T. Sicard, J.L. Kennedy, S. McCullagh, and M.J. Rapoport. 2008. The serotonin transporter polymorphisms and major depression following traumatic brain injury. Brain Injury 22(6): 471-479.

Chee, M.W.L., J.C. Tan, H. Zheng, S. Parimal, D.H. Weissman, V. Zagorodnov, and D.F. Dinges. 2008. Lapsing during sleep deprivation is associated with distributed changes in brain activation. Journal of Neuroscience 28(21): 5519-5528.

Chen, S.C., J.M. Davis, C. Carlstedt, C. Kline, E.A. Murphy, and M.D. Carmichael. 2008. Effects of quercetin on maximal oxygen consumption and fatigue during prolonged exercise. Medicine & Science in Sports & Exercise 40(5): S165.

Cheuvront, S.N., R. Carter III, M.A. Kolka, H.R. Lieberman, M.D. Kellogg, and M.N. Sawka. 2004. Branched-chain amino acid supplementation and human performance when hypohydrated in the heat. Journal of Applied Physiology 97(4): 1275-1282.

Chinevere, T.D., R.D. Sawyer, A.R. Creer, R.K. Conlee, and A.C. Parcell. 2002. Effects of L-tyrosine and carbohydrate ingestion on endurance exercise performance. Journal of Applied Physiology 93(5): 1590-1597.

Coggan, A.R., and E.F. Coyle. 1991. Carbohydrate ingestion during prolonged exercise: Effects on metabolism and performance. Exercise and Sport Sciences Reviews 19: 1-40.

Comalada, M., D. Camuesco, S. Sierra, I. Ballester, J. Xaus, J. Gálvez, and A. Zarzuelo. 2005. *In vivo* quercitrin anti-inflammatory effect involves release of quercetin, which inhibits inflammation through down-regulation of the NF-κB pathway. European Journal of Immunology 35(2): 584-592.

Comalada, M., I. Ballester, E. Bailón, S. Sierra, J. Xaus, J. Gálvez, F.S. de Medina, and A. Zarzuelo. 2006. Inhibition of pro-inflammatory markers in primary bone marrow-derived mouse macrophages by naturally occurring flavonoids: Analysis of the structure–activity relationship. Biochemical Pharmacology 72(8):1010-1021.

Conlay, L.A., L.A. Sabounjian, and R.J. Wurtman. 1992. Exercise and neuromodulators: Choline and acetylcholine in marathon runners. International Journal of Sports Medicine 13(S1): S141-S142.

Cook, D.B., P.J. O'Connor, G. Lange, and J. Steffener. 2007. Functional neuroimaging correlates of mental fatigue induced by cognition among chronic fatigue syndrome patients and controls. Neuroimage 36(1): 108-122.

Cooper, J.R., F.E. Bloom, and R.H. Roth. 2003. The Biochemical Basis of Neuropharmacology, 8th ed. New York, N.Y.: Oxford University Press.

Cotman, C.W., and N.C. Berchtold. 2002. Exercise: A behavioral intervention to enhance brain health and plasticity. Trends in Neurosciences 25(6): 295-301.

Dantzer, R., and K.W. Kelley. 1989. Stress and immunity: An integrated view of relationships between the brain and the immune system. Life Sciences 44(26): 1995-2008.

Davis, J.M. 2000. Nutrition, neurotransmitters and CNS fatigue. Pp. 171-183 in Nutrition in Sport. R.J. Maughan, ed. Malden, Mass.: Blackwell Science.

Davis, J.M., S.P. Bailey, J.A. Woods, F.J. Galiano, M.T. Hamilton, and W.P. Bartoli. 1992. Effects of carbohydrate feedings on plasma free tryptophan and branched-chain amino acids during prolonged cycling. European Journal of Applied Physiology 65(6): 513-519.

Davis, J.M., and S.P. Bailey. 1997. Possible mechanisms of central nervous system fatigue during exercise. Medicine & Science in Sports & Exercise 29(1): 45-57.

Davis, J.M., and R. Fitts. 1998. Mechanisms of muscular fatigue. Pp. 182-188 in ACSM's Resource Manual for Guidelines for Exercise Testing and Prescription, 3rd edition. J.L. Roitman, ed. Baltimore, Md.: Williams & Wilkins.

Davis, J.M., E.A. Murphy, M.D. Carmichael, J.B. Davis, J.D. Gangemi, and E.P. Mayer. 2007a. Effect of quercetin feedings on tissue mitochondrial enzymes and performance in mice. Medicine & Science in Sports & Exercise 39(5) Supplement: S90.

Davis, J.M., E.A. Murphy, M.D. Carmichael, M.R. Zielinski, C.M. Groschwitz, A.S. Brown, J.D. Gangemi, A. Ghaffar, and E.P. Mayer. 2007b. Curcumin effects on inflammation and performance recovery following eccentric exercise-induced muscle damage. American Journal of Physiology: Regulatory, Integrative and Comparative Physiology 292(6): R2168-R2173.

Davis, J.M., R.S. Welsh, K.L. De Volve, and N.A. Alderson. 1999. Effects of branched-chain amino acids and carbohydrate on fatigue during intermittent, high-intensity running. International Journal of Sports Medicine 20(5): 309-314.

Davis, J.M., Z. Zhao, H.S. Stock, K.A. Mehl, J. Buggy, and G.A. Hand. 2003b. Central nervous system effects of caffeine and adenosine on fatigue. American Journal of Physiology: Regulatory, Integrative and Comparative Physiology 284(2): R399-R404.

Davis, M., K. Ressler, B.O. Rothbaum, and R. Richardson. 2006. Effects of D-cycloserine on extinction: Translation from preclinical to clinical work. Biological Psychiatry 60(4): 369-375.

Davis, M., D.L. Walker, and K.M. Myers. 2003a. Role of the amygdala in fear extinction measured with potentiated startle. Annals of the New York Academy of Sciences 985: 218-232.

Deuster, P.A., A. Singh, R. Coll, D.E. Hyde, and W.J. Becker. 2002. Choline ingestion does not modify physical or cognitive performance. Military Medicine 167(12): 1020-1025.

Dias, A.S., M. Porawski, M. Alonso, N. Marroni, P.S. Collado, and J. González-Gallego. 2005. Quercetin decreases oxidative stress, NF-κB activation, and iNOS overexpression in liver of streptozotocin-induced diabetic rats. The Journal of Nutrition 135(10): 2299-2304.

Dworak, M., P. Diel, S. Voss, W. Hollmann, and H.K. Strüder. 2007. Intense exercise increases adenosine concentrations in rat brain: Implications for a homeostatic sleep drive. Neuroscience 150(4): 789-795.

Dwyer, D. 2002. Glucose Metabolism in the Brain. Boston, Mass.: Academic Press.

Elsabagh, S., D.E. Hartley, O. Ali, E.M. Williamson, and S.E. File. 2005. Differential cognitive effects of *Ginkgo biloba* after acute and chronic treatment in healthy young volunteers. Psychopharmacology 179(2): 437-446.

Erlanger, D.M., K.C. Kutner, J.T. Barth, and R. Barnes. 1999. Neuropsychology of sports-related head injury: Dementia pugilistica to post concussion syndrome. Clinical Neuropsychology 13(2): 193-209.

Euston, D.R., M. Tatsuno, and B.L. McNaughton. 2007. Fast-forward playback of recent memory sequences in prefrontal cortex during sleep. Science 318(5853): 1147-1150.

Evans, M., and S.A. Amiel. 1998. Carbohydrates as a cerebral metabolic fuel. Journal of Pediatric Endocrinology and Metabolism 11(supplement 1): 99-102.

Fredholm, B.B., K. Bättig, J. Holmén, A. Nehlig, and E.E. Zvartau. 1999. Actions of caffeine in the brain with special reference to factors that contribute to its widespread use. Pharmacological Reviews 51(1): 83-133.

Gandevia, S.C. 2001. Spinal and supraspinal factors in human muscle fatigue. Physiological Reviews 81(4): 1725-1789.

Garrett, B.E., and R.R. Griffiths. 1997. The role of dopamine in the behavioral effects of caffeine in animals and humans. Pharmacology Biochemistry and Behavior 57(3): 533-541.

Gottlieb, D.J., G.T. O'Connor, and J.B. Wilk. 2007. Genome-wide association of sleep and circadian phenotypes. BMC Medical Genetics 8(Suppl. 1): S9.

Groeger, J.A., A.U. Viola, J.C.Y. Lo, M. von Schantz, S.N. Archer, and D.-J. Dijk. 2008. Early morning executive functioning during sleep deprivation is compromised by a *PERIOD3* polymorphism. SLEEP 31(8): 1159-1167.

Guezennec, C.Y., A. Abdelmalki, B. Seirurier, D. Merino, X. Bigard, M. Berthelot, C. Pierard, and M. Peres. 1998. Effects of prolonged exercise on brain ammonia and amino acids. International Journal of Sports Medicine 19(5): 323-327.

Haglund, M.E.M., P.S. Nestadt, N.S. Cooper, S.M. Southwick, and D.S. Charney. 2007. Psychobiological mechanisms of resilience: Relevance to prevention and treatment of stress-related psychopathology. Development and Psychopathology 19(3): 889-920.

Hampson, D.B., A. St. Clair Gibson, M.I. Lambert, and T.D. Noakes. 2001. The influence of sensory cues on the perception of exertion during exercise and central regulation of exercise performance. Sports Medicine 31(13): 935-952.

Harwood, M., B. Danielewska-Nikiel, J.F. Borzelleca, G.W. Flamm, G.M. Williams, and T.C. Lines. 2007. A critical review of the data related to the safety of quercetin and lack of evidence of in vivo toxicity, including lack of genotoxic/carcinogenic properties. Food and Chemical Toxicology 45(11): 2179-2205.

Haskell, C.F., D.O. Kennedy, K.A. Wesnes, and A.B. Scholey. 2005. Cognitive and mood improvements of caffeine in habitual consumers and habitual non-consumers of caffeine. Psychopharmacology 179(4): 813-825.

Hassmèn, P., E. Blomstrand, B. Ekblom, and E.A. Newsholme. 1994. Branched-chain amino acid supplementation during 30-km competitive run: Mood and cognitive performance. Nutrition 10(5): 405-410.

Heim, C., and C.B. Nemeroff. 2002. Neurobiology of early life stress: Clinical studies. Seminars in Clinical Neuropsychiatry 7(2): 147-159.

Heim, C., D.J. Newport, S. Heit, Y.P. Graham, M. Wilcox, R. Bonsall, A.H. Miller, and C.B. Nemeroff. 2000. Pituitary–adrenal and autonomic responses to stress in women after sexual and physical abuse in childhood. Journal of the American Medical Association 284(5): 592-597.

Heinrichs, M., T. Baumgartner, C. Kirschbaum, and U. Ehlert. 2003. Social support and oxytocin interact to suppress cortisol and subjective responses to psychosocial stress. Biological Psychiatry 54(12): 1389-1398.

Hoge, C.W., D. McGurk, J.L. Thomas, A.L. Cox, C.C. Engel, and C.A. Castro. 2008. Mild traumatic brain injury in U.S. soldiers returning from Iraq. New England Journal of Medicine 358(5): 453-463.

Huang, S.-M., C.-H. Wu, and G.-C. Yen. 2006. Effects of flavonoids on the expression of the pro-inflammatory response in human monocytes induced by ligation of the receptor for AGEs. Molecular Nutrition and Food Research 50(12): 1129-1139.

Huston, J.P., H.L. Haas, F. Boix, M. Pfister, U. Decking, J. Schrader, and R.K.W. Schwarting. 1996. Extracellular adenosine levels in neostriatum and hippocampus during rest and activity periods of rats. Neuroscience 73(1): 99-107.

Jordan, B.D. 2000. Chronic traumatic brain injury associated with boxing. Seminars in Neurology 20(2): 179-185.

Jordan, B.D., N.R. Relkin, L.D. Ravdin, A.R. Jacobs, A. Bennett, and S. Gandy. 1997. Apolipoprotein E epsilon4 associated with chronic traumatic brain injury in boxing. Journal of the American Medical Association 278(2): 136-140.

Kaffman, A., and M.J. Meaney. 2007. Neurodevelopmental sequelae of postnatal maternal care in rodents: Clinical and research implications of molecular insights. Journal of Child Psychology and Psychiatry and Allied Disciplines 48(3-4): 224-244.

Kaufman, J., B.-Z. Yang, H. Douglas-Palumberi, S. Houshyar, D. Lipschitz, J.H. Krystal, and J. Gelernter. 2004. Social supports and serotonin transporter gene moderate depression in maltreated children. Proceedings of the National Academy of Sciences of the United States of America 101(49): 17316-17321.

Kennedy, D.O., A.B. Scholey, and K.A. Wesnes. 2001. Differential, dose dependent changes in cognitive performance following acute administration of a *Ginkgo biloba/Panax ginseng* combination to healthy young volunteers. Nutritional Neuroscience 4(5): 399-412.

Kennedy, D.O., A.B. Scholey, and K.A. Wesnes. 2002. Modulation of cognition and mood following administration of single doses of *Ginkgo biloba,* ginseng, and a ginkgo/ginseng combination to healthy young adults. Physiology and Behavior 75(5): 739-751.

Kennedy, D.O., C.F. Haskell, K.A. Wesnes, and A.B. Scholey. 2004. Improved cognitive performance in human volunteers following administration of guarana (*Paullinia cupana*) extract: Comparison and interaction with *Panax ginseng*. Pharmacology Biochemistry and Behavior 79(3): 401-411.

Kent, S., R.M. Bluthe, R. Dantzer, A.J. Hardwick, K.W. Kelley, N.J. Rothwell, and J.L. Vannice. 1992. Different receptor mechanisms mediate the pyrogenic and behavioral effects of interleukin 1. Proceedings of the National Academy of Sciences of the United States of America 89(19): 9117-9120.

Kilpatrick, D.G., K.C. Koenen, K.J. Ruggiero, R. Acierno, S. Galea, H.S. Resnick, J. Roitzsch, J. Boyle, and J. Gelernter. 2007. The serotonin transporter genotype and social support and moderation of posttraumatic stress disorder and depression in hurricane-exposed adults. American Journal of Psychiatry 164(11): 1693-1699.

Koenigs, M., E.D. Huey, V. Raymont, B. Cheon, J. Solomon, E.M. Wassermann, and J. Grafman. 2008. Focal brain damage protects against post-traumatic stress disorder in combat veterans. Nature Neuroscience 11(2): 232-237.

Koslowski, S., Z. Brzezinska, K. Nazae, and E. Turlejska. 1981. Carbohydrate availability for the brain and muscle as a factor modifying sympathetic activity during exercise in dogs. Pp. 54-62 in Biochemistry of Exercise IV. J. Poortmans and G. Niset, eds. Baltimore, Md.: University Park Press.

Lahl, O., C. Wispel, B. Willigens, and R. Pietrowsky. 2008. An ultra short episode of sleep is sufficient to promote declarative memory performance. Journal of Sleep Research 17(1): 3-10.

Lakhanpal, P., and D.K. Rai. 2007. Quercetin: A versatile flavonoid. Internet Journal of Medical Update 2(2): 20-35.

Lesch, K.P., and L. Gutknecht. 2005. Pharmacogenetics of the serotonin transporter. Progress in Neuro-Psychopharmacology and Biological Psychiatry 29(6): 1062-1073.

Lichtman, J.W., J. Livet, and J.R. Sanes. 2008. A technicolour approach to the connectome. Nature Reviews Neuroscience 9(6): 417-22.

Lieberman, H.R. 2003. Nutrition, brain function and cognitive performance. Appetite 40(3): 245-254.

Lieberman, H.R., C.M. Falco, and S.S. Slade. 2002. Carbohydrate administration during a day of sustained aerobic activity improves vigilance, as assessed by a novel ambulatory monitoring device, and mood. The American Journal of Clinical Nutrition 76(1): 120-127.

Lieberman, H.R., J.H. Georgelis, T.J. Maher, and S.K. Yeghiayan. 2005. Tyrosine prevents effects of hyperthermia on behavior and increases norepinephrine. Physiology and Behavior 84(1): 33-38.

Lim, J., and D.F. Dinges. 2008. Sleep deprivation and vigilant attention. Annals of the New York Academy of Sciences 1129: 305-322.

Lyamin, O.I., P.R. Manger, S.H. Ridgway, L.M. Mukhametov, and J.M. Siegel. 2008. Cetacean sleep: An unusual form of mammalian sleep. Neuroscience and Biobehavioral Reviews 32: 1451-1484.

Magill, R.A., W.F. Waters, G.A. Bray, J. Volaufova, S.R. Smith, H.R. Lieberman, N. McNevin, and D.H. Ryan. 2003. Effects of tyrosine, phentermine, caffeine d-amphetamine, and placebo on cognitive and motor performance deficits during sleep deprivation. Nutritional Neuroscience 6(4): 237-246.

Magkos, F., and S.A. Kavouras. 2004. Caffeine and ephedrine: Physiological, metabolic and performance-enhancing effects. Sports Medicine 34(13): 871-889.

Mahoney, C.R., J. Castellani, F.M. Kramer, A. Young, and H.R. Lieberman. 2007. Tyrosine supplementation mitigates working memory decrements during cold exposure. Physiology and Behavior 92(4): 575-582.

Marshall, L., and J. Born. 2007. The contribution of sleep to hippocampus-dependent memory consolidation. Trends in Cognitive Science 11(10): 442-450.

Martínez-Flórez, S., B. Gutiérrez-Fernández, S. Sánchez-Campos, J. González-Gallego, and M.J. Tuñón. 2005. Quercetin attenuates nuclear factor-κB activation and nitric oxide production in interleukin-1β-activated rat hepatocytes. The Journal of Nutrition 135(6): 1359-1365.

McCormack, K., M.M. Sanchez, M. Bardi, and D. Maestripieri. 2006. Maternal care patterns and behavioral development of rhesus macaque abused infants in the first 6 months of life. Developmental Psychobiology 48(7): 537-550.

McEwen, B.S. 2007. Physiology and neurobiology of stress and adaptation: Central role of the brain. Physiological Reviews 87(3): 873-904.

McNay, E.C., T.M. Fries, and P.E. Gold. 2000. Decreases in rat extracellular hippocampal glucose concentration associated with cognitive demand during a spatial task. Proceedings of the National Academy of Sciences of the United States of America 97(6): 2881-2885.

McNay, E.C., R.C. McCarty, and P.E. Gold. 2001. Fluctuations in brain glucose concentration during behavioral testing: Dissociations between brain areas and between brain and blood. Neurobiology of Learning and Memory 75(3): 325-337.

Meaney, M.J., M. Szyf, and J.R. Seckl. 2007. Epigenetic mechanisms of perinatal programming of hypothalamic-pituitary-adrenal function and health. Trends in Molecular Medicine 13(7): 269-277.

Miller A.H. 2009. Mechanisms of cytokine-induced behavioral changes: Psychoneuroimmunology at the translational interface. Brain, Behavior, and Immunity 23(2): 149-158.

Min, Y.D., C.H. Choi, H. Bark, H.Y. Son, H.H. Park, S. Lee, J.W. Park, E.K. Park, H.I. Shin, and S.H. Kim. 2007. Quercetin inhibits expression of inflammatory cytokines through attenuation of NF-κB and p38 MAPK in HMC-1 human mast cell line. Inflammation Research 56(5): 210-215.

Morgan, C.A., III, S. Southwick, G. Hazlett, A. Rasmusson, G. Hoyt, Z. Zimolo, and D. Charney. 2004. Relationships among plasma dehydroepiandrosterone sulfate and cortisol levels, symptoms of dissociation, and objective performance in humans exposed to acute stress. Archives of General Psychiatry 61(8): 819-825.

Morgan, C.A., III, S. Wang, J. Mason, S.M. Southwick, P. Fox, G. Hazlett, D.S. Charney, and G. Greenfield. 2000. Hormone profiles in humans experiencing military survival training. Biological Psychiatry 47(10): 891-901.

Naidu, P.S., A. Singh, and S.K. Kulkarni, 2004. Reversal of reserpine-induced orofacial dyskinesia and cognitive dysfunction by quercetin. Pharmacology 70(2): 59-67.

NCCAM (National Center for Complementary and Alternative Medicine, National Institutes of Health). 2008. Herbs at a Glance: Turmeric. Available online at http://nccam.nih.gov/health/turmeric/. Last accessed September 5, 2008.

Neumeister, A., X.-Z. Hu, D.A. Luckenbaugh, M. Schwarz, A.C. Nugent, O. Bonne, P. Herscovitch, D. Goldman, W.C. Drevets, and D.S. Charney. 2006. Differential effects of 5-HTTLPR genotypes on the behavioral and neural responses to tryptophan depletion in patients with major depression and controls. Archives of General Psychiatry 63(9): 978-986.

Newsholme, E.A., I.N. Acworth, and E. Blomstrand. 1987. Amino acids, brain neurotransmitters and a functional link between muscle and brain that is important in sustained exercise. Pp. 127-133 in Advances in Myochemistry. B. Genzi, ed. London, England: John Libby Eurotext.

Nieman, D.C., D.A. Henson, J.M. Davis, E.A. Murphy, D.P. Jenkins, S.J. Gross, M.D. Carmichael, J.C. Quindry, C.L. Dumke, A.C. Utter, S.R. McAnulty, L.S. McAnulty, N.T. Triplett, and E.P. Mayer. 2007. Quercetin's influence on exercise-induced changes in plasma cytokines and muscle and leukocyte cytokine mRNA. Journal of Applied Physiology 103(5): 1728-1735.

NRC (National Research Council). 2001. Caffeine for the Sustainment of Mental Task Performance: Formulations for Military Operations. Washington, D.C.: National Academy Press.

Nybo, L. 2003. CNS fatigue and prolonged exercise: Effect of glucose supplementation. Medicine & Science in Sports & Exercise 35(4): 589-594.

Nybo, L. 2008. Hyperthermia and fatigue. Journal of Applied Physiology 104(3): 871-878.

Nybo, L., K. Moller, B.K. Pedersen, B. Nielsen, and N.H. Secher. 2003a. Association between fatigue and failure to preserve cerebral energy turnover during prolonged exercise. Acta Physiologica Scandinavica 179(1): 67-74.

Nybo, L., B. Nielsen, E. Blomstrand, K. Moller, and N. Secher. 2003b. Neurohumoral responses during prolonged exercise in humans. Journal of Applied Physiology 95(3): 1125-1131.

Nybo, L., M.K. Dalsgaard, A. Steensberg, K. Moller, and N.H. Secher. 2005. Cerebral ammonia uptake and accumulation during prolonged exercise in humans. Journal of Physiology 563(1): 285-290.

O'Leary, K.A., S. de Pascual-Tereasa, P.W. Needs, Y.-P. Bao, N.M. O'Brien, and G. Williamson. 2004. Effect of flavonoids and vitamin E on cyclooxygenase-2 (COX-2) transcription. Mutation Research/Fundamental and Molecular Mechanisms of Mutagenesis 551(1-2): 245-254.

Parker, K.J., C.L. Buckmaster, A.F. Schatzberg, and D.M. Lyons. 2004. Prospective investigation of stress inoculation in young monkeys. Archives of General Psychiatry 61(9): 933-941.

Parker, K.J., C.L. Buckmaster, K.R. Justus, A.F. Schatzberg, and D.M. Lyons. 2005. Mild early life stress enhances prefrontal-dependent response inhibition in monkeys. Biological Psychiatry 57(8): 848-855.

Patil, C.S., V.P. Singh, P.S.V. Satyanarayan, N.K. Jain, A. Singh, and S.K. Kulkarni. 2003. Protective effect of flavonoids against aging- and lipopolysaccharide-induced cognitive impairment in mice. Pharmacology 69(2): 59-67.

Patil, S.T., L. Zhang, F. Martenyi, S.L. Lowe, K.A. Jackson, B.V. Andreev, A.S. Avedisova, L.M. Bardenstein, I.Y. Gurovich, M.A. Morozova, S.N. Mosolov, N.G. Neznanov, A.M. Reznik, A.B. Smulevich, V.A. Tochilov, B.G. Johnson, J.A. Monn, and D.D. Schoepp. 2007. Activation of mGlu2/3 receptors as a new approach to treat schizophrenia: A randomized phase 2 clinical trial. Nature Medicine 13(9): 1102-1107.

Paulus, M.P., and M.B. Stein. 2007. Role of functional magnetic resonance imaging in drug discovery. Neuropsychology Review 17(2): 179-188.

Phan, K.L., M. Angstadt, J. Golden, I. Onyewuenyi, A. Popovska, and H. de Wit. 2008. Cannabinoid modulation of amygdala reactivity to social signals of threat in humans. Journal of Neuroscience 28(10): 2313-2319.

Pillay, P., and P.R. Manger. 2004. Testing thermogenesis as the basis for the evolution of cetacean sleep phenomenology. Journal of Sleep Research 13(4): 353-358.

Porkka-Heiskanen, T. 1999. Adenosine in sleep and wakefulness. Annals of Medicine 31(2): 125-129.

Porkka-Heiskanen, T., R.E. Strecker, M. Thakkar, A.A. Bjorkum, R.W. Greene, and R.W. McCarley. 1997. Adenosine: A mediator of the sleep-inducing effects of prolonged wakefulness. Science 276(5316): 1265-1268.

RAND. 2008. Research Highlights—Invisible Wounds: Mental Health and Cognitive Care Needs of America's Returning Veterans. Available at http://www.rand.org/pubs/research_briefs/RB9336/. Last accessed September 5, 2008.

Rauch, S.L., L.M. Shin, and E.A. Phelps. 2006. Neurocircuitry models of posttraumatic stress disorder and extinction: Human neuroimaging research—past, present, and future. Biological Psychiatry 60(4): 376-382.

Reay, J.L., D.O. Kennedy, and A.B. Scholey. 2005. Single doses of *Panax ginseng* (G115) reduce blood glucose levels and improve cognitive performance during sustained mental activity. Journal of Psychopharmacology 19(4): 357-365.

Reay, J.L., D.O. Kennedy, and A.B. Scholey. 2006. Effects of *Panax ginseng*, consumed with and without glucose, on blood glucose levels and cognitive performance during sustained "mentally demanding" tasks. Journal of Psychopharmacology 20(6): 771-781.

Reis, J., O.B. Swayne, Y. Vandermeeren, M. Camus, M.A. Dimyan, M. Harris-Love, M.A. Perez, P. Ragert, J.C. Rothwell, and L.G. Cohen. 2008. Contribution of transcranial magnetic stimulation to the understanding of cortical mechanisms involved in motor control. Journal of Physiology 586(2): 325-351.

Ressler, K.J., B.O. Rothbaum, L. Tannenbaum, P. Anderson, K. Graap, E. Zimand, L. Hodges, and M. Davis. 2004. Cognitive enhancers as adjuncts to psychotherapy: Use of D-cycloserine in phobic individuals to facilitate extinction of fear. Archives of General Psychiatry 61(11): 1136-1144.

Rétey, J.V., M. Adam, E. Honegger, R. Khatami, U.F.O. Luhmann, H.H. Jung, W. Berger, and H.-P. Landolt. 2005. A functional genetic variation of adenosine deaminase affects the duration and intensity of deep sleep in humans. Proceedings of the National Academy of Sciences of the United States of America 102(43): 15676-15681.

Rizzo, A.A., K. Graap, K. Perlman, R.N. McLay, B.O. Rothbaum, G. Reger, T. Parsons, J. Difede, and J. Pair. 2008. Virtual Iraq: Initial results from a VR exposure therapy application for combat-related PTSD. Pp. 420-425 in Medicine Meets Virtual Reality 16—Parallel, Combinatorial, Convergent: NextMed by Design, Volume 132 in Studies in Health Technology and Informatics. J.D. Westwood, R.S. Haluck, H.M. Hoffman, G.T. Mogel, R. Phillips, R.A. Robb, and K.G. Vosburgh, eds. Fairfax, Va.: IOS Press, Inc.

Robinson, R.G., and J.T. Coyle. 1980. The differential effect of right versus left hemispheric cerebral infarction on catecholamines and behavior in the rat. Brain Research 188(1): 63-78.

Robinson, P.J., and S.I. Rapoport. 1986. Glucose transport and metabolism in the brain. American Journal of Physiology: Regulatory, Integrative and Comparative Physiology 250(1): R127-R136.

Roozendaal, B., G.K. Hui, I.R. Hui, D.J. Berlau, J.L. McGaugh, and N.M. Weinberger. 2006. Basolateral amygdala noradrenergic activity mediates corticosterone-induced enhancement of auditory fear conditioning. Neurobiology of Learning and Memory 86(3): 249-255.

Roy, A., X.-Z. Hu, M.N. Janal, and D. Goldman. 2007. Interaction between childhood trauma and serotonin transporter gene variation in suicide. Neuropsychopharmacology 32(9): 2046-2052.

Rutter, M. 1993. Resilience: Some conceptual considerations. Journal of Adolescent Health 14(8): 626-631.

Sapolsky, R.M. 2003. Stress and plasticity in the limbic system. Neurochemical Research 28(11): 1735-1742.

Sayer, N.A., C.E. Chiros, B. Sigford, S. Scott, B. Clothier, T. Pickett, and H.L. Lew. 2008. Characteristics and rehabilitation outcomes among patients with blast and other injuries sustained during the Global War on Terror. Archives of Physical Medicine and Rehabilitation 89(1): 163-170.

Scherrer, J.F., H. Xian, M.J. Lyons, J. Goldberg, S.A. Eisen, W.R. True, M. Tsuang, K.K. Buchholz, and K.C. Koenen. 2008. Posttraumatic stress disorder; combat exposure; and nicotine dependence, alcohol dependence, and major depression in male twins. Comprehensive Psychiatry 49(3): 297-304.

Schneiderman, A.I., E.R. Braver, and H.K. Kang. 2008. Understanding sequelae of injury mechanisms and mild traumatic brain injury incurred during the conflicts in Iraq and Afghanistan: Persistent postconcussive symptoms and posttraumatic stress disorder. American Journal of Epidemiology 167(12): 1446-1452.

Segman, R.H., R. Cooper-Kazaz, F. Macciardi, T. Goltser, Y. Halfon, T. Dobroborski, and A.Y. Shalev. 2002. Association between the dopamine transporter gene and posttraumatic stress disorder. Molecular Psychiatry 7(8): 903-907.

Sharma, V., M. Mishra, S. Ghosh, R. Tewari, A. Basu, P. Seth, and E. Sen. 2007. Modulation of interleukin-1β mediated inflammatory response in human astrocytes by flavonoids: Implications in neuroprotection. Brain Research Bulletin 73(1-3): 55-63.

Sibilia, V., N. Lattuada, D. Rapetti, F. Pagani, D. Vincenza, I. Bulgarelli, V. Locatelli, F. Guidobono, and C. Netti. 2006. Ghrelin inhibits inflammatory pain in rats: Involvement of the opioid system. Neuropharmacology 51(3): 497-505.

Singh, A., P.S. Naidu, and S.K. Kulkarni. 2003. Reversal of aging and chronic ethanol-induced cognitive dysfunction by quercetin, a bioflavonoid. Free Radical Research 37(11): 1245-1252.

Smith, J.L., P.G. Martin, S.C. Gandevia, and J.L. Taylor. 2007. Sustained contraction at very low forces produces prominent supraspinal fatigue in human elbow flexor muscles. Journal of Applied Physiology 103(2): 560-568.

Smith, T.C., M.A.K. Ryan, D.L. Wingard, D.J. Slymen, J.F. Sallis, and D. Kritz-Silverstein, for the Millennium Cohort Study Team. 2008. New onset and persistent symptoms of post-traumatic stress disorder self reported after deployment and combat exposures: Prospective population based US military cohort study. British Medical Journal 336(7640): 366-371.

Smyth, J., M.C. Ockenfels, L. Porter, C. Kirschbaum, D.H. Hellhammer, and A.A. Stone. 1998. Stressors and mood measured on a momentary basis are associated with salivary cortisol secretion. Psychoneuroendocrinology 23(4):353-370.

Spiegel, K., K. Knutson, R. Leproult, E. Tasali, and E. Van Cauter. 2005. Sleep loss: A novel risk factor for insulin resistance and Type 2 diabetes. Journal of Applied Physiology 99(5): 2008-2019.

Stone, A.A., J.E. Schwartz, J. Smyth, C. Kirschbaum, S. Cohen, D. Hellhammer, and S. Grossman. 2001. Individual differences in the diurnal cycle of salivary free cortisol: A replication of flattened cycles for some individuals. Psychoneuroendocrinology 26(3): 295-303.

Stuhmiller, J.H. 2008. Blast injury: Translating research into operational medicine (pre-print). Chapter 10 in Quantitative Physiology: Problems and Concepts in Military Operational Medicine. Textbook of Military Medicine 2008. W.R. Santee and K.E. Friedl, eds. Available online at http://www.bordeninstitute.army.mil/published_volumes/blast_injury/blast_injury.pdf. Last accessed September 8, 2008.

Strüder, H.K., W. Hollman, P. Platen, M. Donike, A. Gotzmann, and K. Weber. 1998. Influence of paroxetine, branched-chain amino acids and tyrosine on neuroendocrine system responses and fatigue in humans. Hormone and Metabolic Research 30(4): 188-194.

Strüder, H.K., W. Hollmann, P. Platen, J. Duperly, H.G. Fischer, and K. Weber. 1996. Alterations in plasma free tryptophan and large neutral amino acids do not affect perceived exertion and prolactin during 90 min of treadmill exercise. International Journal of Sports Medicine 17(2): 73-79.

Tafti, M., and P. Franken. 2007. Molecular analysis of sleep. Cold Spring Harbor Symposium in Quantitative Biology 72: 573-578.

Tasali, E., R. Leproult, D.A. Ehrmann, and E. Van Cauter. 2008. Slow-wave sleep and the risk of type 2 diabetes in humans. Proceedings of the National Academy of Sciences of the United States of America 105(3): 1044-1049.

Taylor, F.B., P. Martin, C. Thompson, J. Williams, T.A. Mellman, C. Gross, E.R. Peskind, and M.A. Raskind. 2008. Prazosin effects on objective sleep measures and clinical symptoms in civilian trauma posttraumatic stress disorder: A placebo-controlled study. Biological Psychiatry 63(6): 629-632.

Toth, C., S. McNeil, and T. Feasby. 2005. Central nervous system injuries in sport and recreation: A systematic review. Sports Medicine 35(8): 685-715.

Van Dongen, H.P.A., M.D. Baynard, G. Maislin, and D.F. Dinges. 2004a. Systematic inter-individual variability differences in neurobehavioral impairment from sleep loss: Evidence of trait-like differential vulnerability. SLEEP 27(3): 423-433.

Van Dongen, H.P.A., G. Maislin, and D.F. Dinges. 2004b. Dealing with inter-individual differences in the temporal dynamics of fatigue and performance: Importance and techniques. Aviation, Space, and Environmental Medicine 75(Suppl. 1): A147-A154.

van Duinen, H., R. Renken, N. Maurits, and I. Zijdewind. 2007. Effects of motor fatigue on human brain activity, an fMRI study. NeuroImage 35(4): 1438-1449.

van Hall, G., J.S. Raaymakers, W.H. Saris, and A.J. Wagenmakers. 1995. Ingestion of branched-chain amino acids and tryptophan during sustained exercise in man: Failure to affect performance. Journal of Physiology 486(Pt. 3): 789-794.

Viola, A.U., S.N. Archer, L.M. James, J.A. Groeger, J.C.Y. Lo, D.J. Skene, M. von Schantz, and D.-J. Dijk. 2007. *PER3* polymorphism predicts sleep structure and waking performance. Current Biology 17(7): 613-618.

Warber, J.P., J.F. Patton, W.J. Tharion, S.H. Zeisel, R.P. Mello, C.P. Kemnitz, and H.R. Lieberman. 2000. The effects of choline supplementation on physical performance. International Journal of Sport Nutrition and Exercise Metabolism 10(2): 170-181.

Welsh, R.S., J.M. Davis, J.R. Burke, and H.G. Williams. 2002. Carbohydrates and physical/mental performance during intermittent exercise to fatigue. Medicine & Science in Sports & Exercise 34(4): 723-731.

Winnick, J.J., J.M. Davis, R.S. Welsh, M.D. Carmichael, E.A. Murphy, and J.A. Blackmon. 2005. Carbohydrate feedings during team sport exercise preserve physical and CNS function. Medicine & Science in Sports & Exercise 37(2): 306-315.

Wray, N.R., C.M. Middeldorp, A.J. Birley, S.D. Gordon, P.F. Sullivan, P.M. Visscher, D.R. Nyholt, G. Willemsen, E.J.C. de Geus, P.E. Slagboom, G.W. Montgomery, N.G. Martin, and D.I. Boomsma. 2008. Genome-wide linkage analysis of multiple measures of neuroticism of 2 large cohorts from Australia and the Netherlands. Archives of General Psychiatry 65(6): 649-658.

Yamamoto, T., and E.A. Newsholme. 2000. Diminished central fatigue by inhibition of the L-system transporter for the uptake of tryptophan. Brain Research Bulletin 52(1): 35-38.

Yehuda, R., and J. LeDoux. 2007. Response variation following trauma: A translational neuroscience approach to understanding PTSD. Neuron 56(1): 19-32.

Zhou, W., D. Xu, X. Peng, Q. Zhang, J. Jia, and K.A. Crutcher. 2008. Meta-analysis of APOE4 allele and outcome after traumatic brain injury. Journal of Neurotrauma 25(4): 279-290.

6

Improving Cognitive and Behavioral Performance

The transformation of the U.S. military services into a highly networked force has markedly increased the need for rapid collection and dissemination of vast amounts of data. This includes the fusion and display of data in formats that can be readily comprehended by soldiers who can then take the appropriate actions.

The combination of optimal performance, data comprehension, and decision making required for this transformation to a networked force comes as the Army evolves into a growing variety of combat platforms. Depending on their specialties, soldiers can be expected to operate equipment ranging from 70-ton Abrams tanks and 35-ton Stryker vehicles to sophisticated manned and unmanned ground and air systems of the Future Combat Systems and to conduct basic dismounted soldier operations in all environments, including urban terrain. All of these operations rely on the soldier maintaining situational awareness and sharing a common operating picture of the battlefield.

The soldiers conducting these varied operations will also contend with stressors on cognitive performance as described in Chapters 4 and 5. The need to remain vigilant and on task continuously for extended periods (longer than 36 hours) in extreme environments (for example, in closed vehicles, where the temperature can exceed 110°F) while enduring the stresses of sustained combat or of security and stability operations will challenge the baseline cognitive and behavioral capabilities of soldiers, who must assimilate and react appropriately to the flow of task-relevant information. In short, each soldier's cognitive performance on his assigned tasks will more than ever before be critical to his or her operational performance.

The recent breakthroughs in neuroimaging and other technologies described in Chapters 2 through 5 allow quantifying the physiological metrics of human attentiveness, cognitive performance, and neural functioning. The knowledge gained is guiding the development of countermeasures against such stressors as fatigue, sleep deprivation, information overload, dehydration and other metabolic stresses, even overtraining. At the same time, many of these techniques appear to offer possibilities for enhancing soldiers' performance beyond their normal, or unaided, baseline capabilities. This chapter assesses the current status and emerging prospects for such neuroscience-informed enhancements.

HOURS OF BOREDOM AND MOMENTS OF TERROR

Many of the tasks that make up a military deployment, especially for operations in a combat theater, can be characterized as "hours of boredom and moments of terror" (Hancock, 1997). During the long periods of waiting that lead up to a combat operation where hostile action may await, the main demand on individual performance is for vigilance and sustained attention (Warm, 1984). The strong association between the psychophysical dimensions of the vigilance task and certain mental workload measures made possible by advances in neuroscience can aid the soldier (Warm et al., 1996). Failures in vigilance can lead to calamity in military operations (Rochlin, 1991; Snook, 2000) as well as in other facets of human life (Hill and Rothblum, 1994; Evan and Manion, 2002). The classic military example is standing watch, where continuous sustained attention is essential but the probability of a threat event at any particular time is low.

Many prominent cases of military failure associated with human error have involved failures of sustained attention (e.g., Miller and Shattuck, 2004). Looked at in terms of classic signal detection theory, such failures are classified as either a false alarm or a miss. Both forms of inappropriate response are problematic, but missing a critical signal for response can result in injury and fatalities not just for the soldier but also for the immediate unit or even beyond. Thus, finding ways to extend attentiveness could have a significant return for overall military performance.

Fortunately, substantial progress has recently been made on the problem of sustained attention. For example, Tripp and Warm (2007) have linked variations in blood flow and blood oxygenation, as measured by transcranial Doppler sonography, with occasions on which observers miss signals. In addition to measuring blood flow and blood oxygenation, which are indirect indicators of neural functioning, event-related potentials may be another way to learn when an individual has missed a critical signal. As discussed in the following section on neuroergonomics, if the data stream of the original event-related potentials is formatted so as to elicit, for example, a P300 response when a miss occurs, then an augmented perception system could be triggered by such an electrophysiological signal. Such techniques for augmenting perception—in this case to improve awareness of a signal—depend on vigilance for catching a specific signal. The interface employed for this task may need to be structured to make best use of the augmentation opportunity, and such designs are a challenge to scientists in the human factors and ergonomics communities (Hancock and Szalma, 2003a, 2003b).

Despite the challenges, work on military applications for this kind of brain-signal-augmented recognition is going forward, as illustrated by two current Defense Advanced Research Projects Agency (DARPA) programs.[1] The Neuroscience for Intelligence Analysts system uses electroencephalography (EEG) to detect a brain signal corresponding to perceptual recognition (which can occur below the level of conscious attention) of a feature of interest in remote (airborne or space-based) imagery. In macaque monkeys, an EEG signature from electrophysiological recordings has been successfully detected for target image presentation rates of up to 72 images per second (Keysers et al., 2001). In the Phase 1 proof-of-concept demonstration of a triage approach to selecting images for closer review, actual intelligence analysts working on a realistic broad-area search task achieved a better than 300 percent improvement in throughput and detection relative to the current standard for operational analysis. There is evidence that this technology can detect at least some classes of unconscious attention, providing support for the notion that perception is not only or always consciously perceived.

The second DARPA program, the Cognitive Technology Threat Warning System, uses a signal-processing system coupled with a helmet-mounted EEG device to monitor brain activity to augment a human sentinel's ability to detect a potential threat image anywhere in a wide field-of-view image seen through a pair of binoculars. Again, the objective is to identify potential features of interest using the brain signal, then warn the soldier-sentinel and direct his or her attention to those features.

If augmentation of signal awareness can enhance performance in continuous-vigilance tasks during the hours of boredom, as illustrated by these DARPA demonstration-experiments, are there opportunities to enhance soldier performance during the infrequent but intense moments of terror? In the modern Army environment, such contexts typically involve surging information loads on individuals who must process all the relevant information quickly and appropriately to avoid the twin performance faults: failure to respond or incorrect response. When peak demands are coming from multiple cognitive tasks—e.g., perceptual judgment, information assimilation to cognitive schema, and choice selection (decision making in a broad sense), all of which must be carried out with urgency—cognitive overload is likely to degrade performance.

As an example, consider a mounted soldier-operator who is monitoring his own formation of manned and unmanned ground vehicles, along with attached unmanned aerial vehicle assets, and is receiving and sending communications over his tactical radio system. At the same time that he notices some problem with one of the unmanned ground vehicles, he loses contact with one of the aerial vehicles and receives preliminary indications of an enemy position on his flank. The soldier in this or analogous circumstances may well have trained for such events individually, but all three occurring simultaneously is likely to produce cognitive overload.

The primary way in which neuroscience can help an individual deal with cognitive overload is through improved methods for load-shedding as the workload stress on the individual increases beyond a manageable level. In effect, the aiding system removes or lessens one or more of the stacked processing-and-response demands on the individual. The load-shedding process can continue as cognitive tasks are sequentially removed. Thus, in the example above, the soldier-operator can focus on the most serious threat—the signs of hostile activity—while his load-shedding system automatically moves into problem-management routines for the two "straying" unmanned vehicles and cuts the incoming message traffic on the radio to just the highest priority messages. Various forms of discrete task allocation have been around in concept since the mid-1950s and in practice since the later 1980s. However, in these existing forms, the aiding system does not receive input on how close the aided individual is to cognitive overload. This particular aspect—monitoring the status of the individual for measures of cognitive overload—is where neuroscience and its technologies for assessing neurophysiological state can contribute to enhancing performance in the moments of terror. In our mounted soldier-operator example, an information workload monitoring system would detect the soldier's nascent cognitive overload condition and activate the automated problem management routines for his straying assets and the radio "hush-down."

In the past decade, much effort has gone into the assessment of neurophysiological indicators of incipient overload. At the forefront of these efforts was the Augmented Cognition (AugCog) program of DARPA. The AugCog objective was to use a number of neural state indicators to control adaptive human–machine interfaces to information systems.

[1]Amy A. Kruse, Defense Sciences Office, DARPA, Briefing to the committee on June 30, 2008.

The neural state indicators were used to assess cognitive overload stress, and when the stress became too great, they would trigger the dynamic load-shedding activity of an interface management system (McBride and Schmorrow, 2005; Schmorrow and Reeves, 2007). This pioneering effort in information workload management via physiological and neural feedback, which is discussed further in Chapter 7 and in greater detail in Appendix D, met some degree of success. It also provides important lessons on the challenges for implementing this type of adaptive aiding technology.

The concept of adaptive aiding, which was first advanced by Rouse (1975) for the U.S. Air Force, builds on a long tradition of behavioral adaptation to environmental constraints (Hancock and Chignell, 1987). However, in a neural-indicator-driven implementation, such as in the original AugCog vision, the adaptation is not managed by the individual alone but is augmented by the aiding system's assessment of the individual's level of cognitive stress or other electropsychological parameters. Research projects in adaptive aiding have focused on systems for such real-world tasks as air traffic control (Hillburn et al., 1997), control of unmanned aerial vehicles (see, for example, Mouloua et al., 2003; Taylor, 2006), and augmentation of fine motor skills such as laparoscopic surgery (Krupa et al., 2002).

In fact, most tasks in which humans perform knowledge-intensive work in conjunction with a complex information management and computational system could probably be improved by better diagnostic representations of the state of the human operator. Ultimately, the question becomes how action is integrated within the brain itself. For example, Minsky (1986) suggested that the brain could be viewed as a "society of mind." In this view, a person's conscious experience is an emergent property arising from the interaction of many cortical subsystems. The way in which these subsystems appear to interact seamlessly[2] may well represent a template for advanced human–machine systems whose goal would be to reproduce the apparent effortlessness with which a person willfully controls his or her own limbs. A sad reality is that we no doubt will learn more about how this interaction works within the brain by working with individuals damaged by war. See, for example, the section of Chapter 7 entitled "Optimal Control Strategies for Brain–Machine Interfaces." The prospect of returning the wounded to their previous level of physiological capability is a potential source of satisfaction. However, the next step beyond recovery of capability—providing capabilities that exceed the norm of natural abilities—would raise ethical issues if, indeed, it became technically feasible (Hancock, 2003).

[2]There may be more cognitive dissonance, or contention and conflict, occurring in such situations than either the operator or an observer of the operator's behavior can detect unaided by information on the operator's neural state (Hancock, in press).

NEUROERGONOMICS

Neuroergonomics has been defined by the individual who coined the term as "the study of the brain and behavior at work" (Parasuraman and Rizzo, 2007). It is one facet, or formalized expression, of the broader field of brain–machine (or sometimes mind–machine) interfaces (Levine et al., 2000; Lebedev and Nicolelis, 2006). Much of the broader field, as discussed in Chapters 5 and 7, has focused on ways to restore full functioning to individuals who have lost limbs or who have suffered some form of cognitive deficit following concussive or kinetic injuries. Although many of the advances in knowledge and in technology for these medical applications are inherently important to all application areas for brain–machine interfaces, the focus of this section is not medical prostheses. Rather, a typical application in neuroergonomics is concerned with enhancing selected capabilities beyond an unaided level, whether or not the individual aided by the system has experienced some degradation in capability. Neuroergonomics deals as much with performance improvement and performance enhancement as with performance recovery. The brief summary below examines some of the opportunities envisioned by those working in this field, as well as some of the barriers, acknowledged and implied, to successful realization of these opportunities.

Specificity of Brain Signals as Control Inputs to a Brain–Machine Interface

For a nonexpert, the advances in neuroscience described in the popular press—and sometimes in proposals seeking funding—can easily be interpreted in ways that overstate the specificity of the signal patterns within the brain that can be monitored with current techniques. Thus, lay individuals frequently ask whether current diagnostic techniques allow an observer to know what the person being observed is thinking. A similarly unrealistic flight of fantasy is that the weapons system of an advanced aircraft can be controlled by thinking in the language of the aircraft's designers or pilots. In general, an expectation that higher levels of cognition can be immediately comprehended by assessing a small number of neural signals is destined for disappointment.

However, the confluence of insights from neuroscience and improvements in complex systems control functions will provide limited opportunities for sending discrete control signals directly to an external system. We can, to some degree, elicit and subsequently measure, with a fair degree of accuracy, discrete responses from the brain. A prime example is the P300 wave, which not only is a potential index of cognitive workload but also can be employed as a form of binary (yes/no) control to a hybrid human–machine monitoring system. Although it remains difficult to distill the P300 wave on a single trial, the signal-to-noise ratio is constantly being improved, as it is for other neurophysiological and allied techniques in neural monitoring. Currently, various forms of brain function can be monitored for use as binary control signals, and simple forms of such controls have been created.

A Pragmatic Approach to Neuroergonomics Applications

Recently, Parasuraman and Wilson (2008) drew a distinction between techniques that measure cerebral metabolic processes, such as transcranial Doppler sonography or functional magnetic resonance imaging (fMRI), and techniques that measure neural activity (neural signaling) per se, such as EEG and event-related potentials. Their primary concern in making this distinction relates to the use of the output from these neurophysiolgical monitoring techniques as inputs to adaptive control systems (Hancock, 2007a, 2007b). Parasuraman and Wilson also considered how sequential improvements in spatial and temporal resolution of these electrophysiolgical measures can provide opportunities for increasingly refined control inputs and thus for increasingly sophisticated control of complex technologies. The eventual goal, whether implied or stated, is to translate an intention to act into a real action—in more colloquial terms, controlling our tools directly with our minds.

The temptation here is to attempt to decide this issue in terms of current and proposed neuroscience methodologies—that is, framing the discussion in terms of what we can measure now and may be able to measure in the near future—and asking how such measures might be used as control signals to a compliant external system (that is, as an input to the defined control interface for the external system). A more practical approach is to ask what the Army and, by extension, its soldiers are expected to do, then consider how these tasks could be accomplished by soldiers interacting with systems via interfaces supported by advanced neuroscience techniques. The discussions of brain–machine interface technologies in Chapter 7 follow this more pragmatic approach.

LEVERAGING EXTERNAL RESEARCH TO ENHANCE SOLDIER PERFORMANCE

This section describes two areas of research on performance enhancement in nonmilitary contexts that have sufficient relevance to Army applications to bear continued monitoring. In addition to discussing these applied research programs, Chapter 7 discusses investments by nonmilitary entities in neuroscience-related technology development. In many cases, those technology opportunities also aim to enhance cognitive and behavioral performance.

Driver Workload Research

Driving a vehicle on highways and streets shared with other vehicles is an integrated, multiple-task behavior that requires proficient performance of different but interrelated skills. These skills rely on interconnected visual, motor, and cognitive brain systems. This common civilian activity of highway driving is regularly performed under conditions of varying task workload and stress. Extensive behavioral research on enhancing driver proficiency has been funded by the private sector—principally the automotive original equipment manufacturers (OEMs)—and by the U.S. Department of Transportation (DOT). These behavioral studies are beginning to be extended and deepened into neurocognitive analyses through complementary research that uses neuroimaging techniques in a laboratory setting. These studies, which are looking at advanced forms of smart vehicles and driver-support information systems, are relevant to soldier performance in high-workload conditions for either combat or noncombat operations that call for combinations of visual, motor, and cognitive processing in the road-driving context. In addition, the models, simulation techniques, and analysis procedures developed for driver workload research and smart-vehicle technology have direct application to human–machine interfaces common in military vehicles, complex weapons systems, and battlefield operations generally.

Neuroscience research has shown the strong influence of key brain communication networks in the degradation of performance in overlearned skills. The methods of neuroscience are very useful for improving the ability to predict behavior by paying attention to the role of identified brain networks. The goal in much of the advanced driver workload/smart-vehicle research is to develop models that predict behavior rather than obtain the most accurate simulation. (Simulation systems for operator training have a different goal, and for them, realism is more important.)

OEM precompetitive collaborative research projects of interest to the Army include the workload measurement aspects of the final reports from the Crash Avoidance Metrics Partnership, as well as the Safety Vehicle Using Adaptive Interface Technology (SAVE-IT) program sponsored by DOT's Research and Technology Innovation Administration (DOT, 2008b). SAVE-IT deals with adaptive interfaces for high-workload environments. The Integrated Vehicle-Based Safety Systems program, also sponsored by the Research and Technology Innovation Administration, will include a study of driver performance in an environment with multiple warning systems that are intended to prevent rear-end, run-off-road, and lane-change crashes (DOT, 2008a; UMTRI, 2008). These investments in driver safety technology have been motivated by an interest in active safety systems to avoid a crash rather than survive one after it happens.

These behavioral studies of workload metrics form the basis for a small set of brain imaging studies in simulated environments. Uchiyama et al. (2003) showed that brain networks are activated in driving-like scenarios in laboratory environments. Young et al. (2006) and Graydon et al. (2004) reported fMRI and magnetoencephalography results showing that a static driving paradigm in a laboratory setting activated the brain network more than did the sum of all the component tasks in the paradigm. They interpreted these results as suggesting that a critical mass of stimulation cues in a laboratory imaging environment can reasonably replicate a real-world scenario for studying driving behavior. Spiers and Maguire (2007) developed a technique for analyzing blocks of driving activity using fMRI and a video game stimulus. Bruns et al. (2005) have used EEG to monitor an individual driving a military vehicle, and Harada et al. (2007) have demonstrated near-infrared spectroscopy technology to monitor the cortical blood flow of an individual operating a civilian automobile.

NASA Neuroscience Research

The National Aeronautics and Space Administration (NASA) has made the second largest federal investment, after DOD, in studying performance under stressful conditions. The National Space and Biomedical Research Institute (NSBRI), a NASA-funded consortium of institutions studying health risks related to long-duration spaceflight, has sought to develop countermeasures to the physical and psychological challenges of spaceflight. The NSBRI also works on technologies to provide medical monitoring and diagnosis capabilities in extreme environments, including cognitive capabilities. NSBRI investigators come from over 70 U.S.-based universities, and the institute is governed by an oversight committee comprising a dozen of its member institutions (NSBRI, 2008a).

The Neurobehavioral and Psychosocial Factors team at the NSBRI seeks to identify neurobehavioral and psychosocial risks to the health, safety, and productivity of space crews. Additional research focuses on developing novel methods of monitoring brain function and behavior and measures that enhance the quality of life for astronauts, along with improving their performance and motivation. This team's current projects range from researching ways to enhance the performance of a team carrying out a space exploration mission to developing new techniques for monitoring cognitive changes and the effects of stress on performance. The team is also developing a computer system that monitors speech patterns for use in predicting changes in mental capacities, such as cognition and decision making, which may be affected by the heightened exposures to radiation or hypoxia that may be encountered on an extended mission. Another current project is to develop a computer-based system for the recognition and treatment of depression (NSBRI, 2008b).

Among the studies completed by the Neurobehavioral and Psychosocial Factors team, Shephard and Kosslyn (2005) developed a portable system to assess nine cognitive functions, including problem solving, attention, and working memory, to provide an early warning sign of stress-related deficits. Dinges et al. (2005, 2007) developed a system using optical computer recognition to track changes in facial expression of astronauts on long spaceflights, when such changes may indicate increased stress.

A second NSBRI team, the Human Factors and Performance team, is studying ways to improve daily living and keep crew members healthy, productive, and comfortable during extended space exploration missions. Overall aims of this team are to reduce performance errors by studying environmental and behavioral factors that could threaten mission success. The team develops information tools to support crew performance and guidelines for human systems design.

Team members are examining ways to improve sleep and scheduling of work shifts and looking at how lighting can improve alertness and performance. Other projects address nutritional countermeasures and how factors in the environment, such as lunar dust, can impact crew health. Recent projects of the Human Factors and Performance team includes research on sleep disruption in space and finding a nutritional counterbalance for the loss of muscle mass and function attributed to long spaceflights. Because rapidly changing light-dark cycles in space can affect the human body's natural circadian cycle, Lockley et al. (2006) have been investigating whether exposure to short-wavelength blue light can be an effective means of shifting the circadian pacemaker, suppressing melatonin, and essentially increasing alertness. Gronfier et al. (2007) found that a modulated light exposure, with bright light pulses of 100 lux being supplied in the evening, can retrain human subjects to a light-dark cycle.

NEUROPHARMACEUTICAL APPROACHES TO PERFORMANCE ENHANCEMENT

Chapter 5 discusses nutritional supplements and pharmaceuticals used to sustain performance (measures to counter environmental stressors) as opposed to enhancing it above an individual's baseline optimum. The committee has significant concerns about the potential for inappropriate use of currently available performance-enhancing drugs by the military.

The caveats noted in Chapter 5 to the off-label use of neuropharmaceuticals to sustain performance, outside the FDA-approved medical indications for prescribing them, apply even more stringently when the intent is to enhance performance beyond the baseline capability. The requirements for specificity and selectivity must be set high and must be clearly met with scientifically sound evidence. And the risk of undesirable and still-unknown side effects must be weighed carefully against any performance benefit using tools to measure the performance improvement and clinical measures to assess the overall effects of the intervention. Such tools may need to be developed. Despite these concerns, it may be worthwhile to continue research on the use of pharmacological agents to optimize performance if the benefits to unique military circumstances clearly outweigh the risks. Future studies may discover enhancers with more striking effects then those currently available (Narkar et al., 2008).

Neuropharmaceuticals might also be applied to influence adversary behavior and decision making. Because pharmaceuticals can no doubt modulate the neurophysiological underpinnings of behavior and performance, they can in principle be used to weaken or incapacitate an adversary, just as they can be used to sustain and strengthen our own soldiers. Although this might be a direction for long-term research, it would also raise substantial ethical, legal (from the perspectives of both U.S. and international law), and strategic issues that should be addressed before the Army supports any such research and before assessing the relevance for Army applications of any non-Army research in this area. As with chemical and biological weapons, the most relevant opportunity for the counteradversary use of pharmaceuticals may be in developing the means to protect our soldiers (and civilians) against pharmacological weapons used against us.

REFERENCES

Bruns, A., K. Hagemann, M. Schrauf, J. Kohmorgen, M. Braun, G. Dornhege, K. Muller, C. Forsythe, K. Dixon, C.E. Lippitt, C.D. Balaban, and W.E. Kincses. 2005. EEG- and context-based cognitive-state classifications lead to improved cognitive performance while driving. P. 1065 in Foundations of Augmented Cognition, Proceedings of the 2005 Human-Computer Interaction Conference. D.D. Schmorrow, ed. Boca Raton, Fla.: CRC Press.

Dinges, D.F., R.L. Rider, J. Dorrian, E.L. McGlinchey, N.L. Rogers, Z. Cizman, S.K. Goldenstein, C. Vogler, S. Venkataraman, and D.N. Metaxas. 2005. Optical computer recognition of facial expressions associated with stress induced by performance demands. Aviation, Space, and Environmental Medicine 76(Supplement 1): B172-B182.

Dinges, D.F., S. Venkataraman, E.L. McGlinchey, and D.N. Metaxas. 2007. Monitoring of facial stress during space flight: Optical computer recognition combining discriminative and generative methods. Acta Astronautica 60(4-7): 341-350.

DOT (U.S. Department of Transportation). 2008a. Integrated vehicle-based safety systems. Available at http://www.its.dot.gov/ivbss/. Last accessed September 8, 2008.

DOT. 2008b. Safety vehicle using adaptive interface technology (SAVE-IT). Available at http://www.volpe.dot.gov/hf/roadway/saveit/index.html. Last accessed September 8, 2008.

Evan, W.M., and M. Manion. 2002. Minding the Machines: Preventing Technological Disasters. Upper Saddle River, N.J.: Prentice Hall.

Graydon, F.X., R. Young, M.D. Benton, R.J. Genik II, S. Posse, L. Hsieh, and C. Green. 2004. Visual event detection during simulated driving: Identifying the neural correlates with functional neuroimaging. Transportation Research Part F: Traffic Psychology and Behaviour 7(4-5): 271-286.

Gronfier, C., K.P. Wright, Jr., R.E. Kronauer, and C.A. Czeisler. 2007. Entrainment of the human circadian pacemaker to longer-than-24-h days. Proceedings of the National Academy of Sciences of the United States of America 104(21): 9081-9086.

Hancock, P.A. 1997. Hours of boredom, moments of terror—or months of monotony, milliseconds of mayhem. Paper presented at the Ninth International Symposium on Aviation Psychology, Columbus, Ohio. April 27-May 1.

Hancock, P.A. 2003. The ergonomics of torture: The moral dimension of evolving human-machine technology. Pp. 1009-1011 in Proceedings of the Human Factors and Ergonomics Society 47. Santa Monica, Calif.: Human Factors and Ergonomics Society.

Hancock, P.A. 2007a. Procedure and dynamic display relocation on performance in a multitask environment. IEEE Transactions on Systems, Man, and Cybernetics—Part A: Systems and Humans 37(1): 47-57.

Hancock, P.A. 2007b. On the process of automation transition in multitask human–machine systems. IEEE Transactions on Systems, Man, and Cybernetics—Part A: Systems and Humans 37(4): 586-598.

Hancock, P.A. In press. The battle for time in the brain. In Time, Limits and Constraints: The Study of Time XIII. J.A. Parker, P.A. Harris, and C. Steineck, eds. Leiden, The Netherlands: Brill.

Hancock, P.A., and M.H. Chignell. 1987. Adaptive control in human–machine systems. Pp. 305-345 in Human Factors Psychology. P.A. Hancock, ed. Amsterdam, The Netherlands: North Holland.

Hancock, P.A., and J.L. Szalma. 2003a. The future of neuroergonomics. Theoretical Issues in Ergonomic Science 4(1-2): 238-249.

Hancock, P.A., and J.L. Szalma. 2003b. Operator stress and display design. Ergonomics in Design 11(2): 13-18.

Harada, H., H. Nashihara, K. Morozumi, H. Ota, and E.A. Hatakeyama. 2007. A comparison of cerebral activity in the prefrontal region between young adults and the elderly while driving. Journal of Physiological Anthropology 26(3): 409-414.

Hill, S.G., and A.M. Rothblum. 1994. Human factors issues in the maritime industry. Proceedings of the Human Factors and Ergonomics Society 38: 862.

Hillburn, B., P.G. Jorna, E.A. Byrne, and R. Parasuraman. 1997. The effect of adaptive air traffic control (ATC) decision aiding on controller mental workload. Pp. 84-91 in Human–Automation Interaction: Research and Practice. M. Mouloua and J. Koonce, eds. Mahwah, N.J.: Lawrence Erlbaum Associates.

Keysers, C., D.K. Xiao, P. Földiák, and D.I. Perrett. 2001. The speed of sight. Journal of Cognitive Neuroscience 13(1): 90-101.

Krupa, A., M. de Mathelin, C. Doignon, J. Gangloff, G. Morel, L. Soler, J. Leroy, and J. Marescaux. 2002. Automatic 3-D positioning of surgical instruments during robotized laparoscopic surgery using automatic visual feedback. Pp. 9-16 in Lecture Notes in Computer Science, Volume 2488: Medical Image Computing and Computer-Assisted Intervention—MICCAI 2002. Berlin: Springer.

Lebedev, M.A., and M.A.L. Nicolelis. 2006. Brain–machine interfaces: Past, present and future. Trends in Neuroscience 29(9): 536-546.

Levine, S.P., J.E. Huggins, S.L. BeMent, R.K. Kushwaha, L.A. Schuh, M.M. Rohde, E.A. Passaro, D.A. Ross, K.V. Elisevich, and B.J. Smith. 2000. A direct brain interface based on event-related potentials. IEEE Transactions on Rehabilitation Engineering 8(2): 180-185.

Lockley, S.W., E.E. Evans, F.A.J.L. Scheer, G.C. Brainard, C.A. Czeisler, and D. Aeschbach. 2006. Short-wavelength sensitivity for the direct effects of light on alertness, vigilance, and the waking electroencephalogram in humans. Sleep 29(2): 161-168.

McBride, D.K., and D.D. Schmorrow, eds. 2005. Quantifying Human Information Processing. Lanham, Md.: Lexington Books.

Miller, N.L., and L.G. Shattuck. 2004. A process model of situated cognition in military command and control. Paper presented at the 2004 Command and Control Research and Technology Symposium, San Diego, Calif.

Minsky, M. 1986. The Society of Mind. New York, N.Y.: Simon and Schuster.

Mouloua, M., R. Gilson, and P.A. Hancock. 2003. Human-centered design of unmanned aerial vehicles. Ergonomics in Design 11(1): 6-11.

Narkar, V.A., M. Downes, R.T. Yu, E. Embler, Y.-X. Wang, E. Banayo, M.M. Mihaylova, M.C. Nelson, Y. Zou, H. Juguilon, H. Kang, R.J. Shaw, and R.M. Evans. 2008. AMPK and PPAR agonists are exercise mimetics. Cell 134(3): 405-415.

NSBRI (National Space and Biomedical Research Institute). 2008a. About NSBRI. Fact sheet. Available at http://www.nsbri.org/About/FactSheet.html. Last accessed September 8, 2008.

NSBRI. 2008b. Neurobehavioral and psychosocial factors: Current team projects. Available at http://www.nsbri.org/Research/Projects/listprojects.epl?team=psycho. Last accessed September 8, 2008.

Parasuraman, R., and M. Rizzo, eds. 2007. Neuroergonomics: The Brain at Work. New York, N.Y.: Oxford University Press.

Parasuraman, R., and G.F. Wilson. 2008. Putting the brain to work: Neuroergonomics past, present and future. Human Factors: Journal of the Human Factors and Ergonomics Society 50(3): 468-474.

Rochlin, G. 1991. Iran Air flight 655: Complex, large-scale military systems and the failure of control. Pp. 95-121 in Social Responses to Large Technical Systems: Control or Anticipation. T.R. La Porte, ed. Boston, Mass.: Kluwer Academic Publishers.

Rouse, W.B. 1975. Human interaction with an intelligent computer in multi-task situations. Pp. 130-143 in Proceedings of the 11th Annual Conference on Manual Control. Seattle, Wash.: Boeing Commercial Airplane Company.

Schmorrow, D.D., and L.M. Reeves, eds. 2007. Foundations of Augmented Cognition. New York, N.Y.: Springer.

Shephard, J.M., and S.M. Kosslyn. 2005. The Minicog Rapid Assessment Battery: Developing a "blood pressure cuff for the mind." Aviation, Space, and Environmental Medicine 76(Supplement 1): B192-B197.

Snook, S.A. 2000. Friendly Fire: The Accidental Shootdown of U.S. Black Hawks over Northern Iraq. Princeton, N.J.: Princeton University Press.

Spiers, H.J., and E.A. Maguire. 2007. Neural substrates of driving behaviour. NeuroImage 36(1): 245-255.

Taylor, R.M. 2006. Human automation integration for supervisory control of UAVs. Pp. 12:1-12:10 in Virtual Media for Military Applications, Meeting Proceedings RTO-MP-HFM-136. Neuilly-sur-Seine, France: RTO.

Tripp, L.D., and J.S. Warm. 2007. Transcranial Doppler sonography. Pp. 82-94 in Neuroergonomics: The Brain at Work. R. Parasuraman and M. Rizzo, eds. New York, N.Y.: Oxford University Press.

Uchiyama, Y., K. Ebe, A. Kozato, T. Okada, and N. Sadato. 2003. The neural substrates of driving at a safe distance: A functional MRI study. Neuroscience Letters 352(3): 199-202.

UMTRI (University of Michigan Transportation Research Institute). 2008. Integrated Vehicle-Based Safety Systems (IVBSS): Phase I Interim Report. Report No. DOT HS 810 952. Available at http://www.nhtsa.dot.gov/staticfiles/DOT/NHTSA/NRD/Multimedia/PDFs/Crash%20Avoidance/2008/810952Lo.pdf. Last accessed September 8, 2008.

Warm, J.S., ed. 1984. Sustained Attention in Human Performance. New York, N.Y.: Wiley.

Warm, J.S., W.N. Dember, and P.A. Hancock. 1996. Vigilance and workload in automated systems. Pp.183-200 in Automation and Human Performance: Theory and Applications. R. Parasuraman and M. Mouloua, eds. Mahwah, N.J.: Lawrence Erlbaum Associates.

Young, R.A., L. Hsieh, F.X. Graydon, R. Genik, M.D. Benton, C.C. Green, S.M. Bowyer, J.E. Moran, and N. Tepley. 2006. Mind-on-the-drive: Real-time functional neuroimaging of cognitive brain mechanisms underlying driver performance and distraction. SAE Transactions 114(6): 454-472.

7

Neuroscience Technology Opportunities

Chapters 3 through 6 discussed neuroscience research leading to developments in key application areas of training, decision making, and performance, including recommendations on predominantly research-based opportunities. This chapter discusses high-risk, high-payoff opportunities for using neuroscience technologies in Army applications. Where the committee identified significant investments in a technology opportunity from nonmilitary federal, commercial, or foreign sectors, the leveraging opportunities for the Army are noted. A section on barriers to Army use of these technology opportunities describes important scientific and technical barriers. The concluding section presents the committee's priorities—"high priority," "priority," and "possible future opportunities"—for Army investment.

A section on technology trends discusses several trends that the committee believes will endure and even grow in significance for Army applications. The Army should establish a mechanism to monitor these trends effectively and have a capability in place to evaluate the applicability of any resulting technology to Army needs and opportunities.

This chapter discusses many technologies and plausible areas of neuroscience research, including experiments that may be facilitated through the use of human subjects. The committee assumes that all such research will be conducted in accordance with the guidelines established in the Belmont Report and subsequent regulations issued by the Office of Human Research Protections of the U.S. Department of Health and Human Services.

The report places technologies in two categories: those that result in "mission-enabling" instruments and those that result in "research-enabling" instruments. In some instances, a technology has applications in both categories. The word "instrument" is used in the most general sense: it could be a pen-and-paper personality inventory, a software-controlled skills survey, a reaction-time analysis method for training assessment, a control interlock system to distribute information among different vehicle crew based on their current workload and baseline cognitive capability, an in-helmet device designed to monitor neural activity or cerebral blood flow, or an advance in imaging technology. Both categories of technology share the common characteristic that neuroscience research, as defined in Chapter 2, plays a key role in their development.

Deployable instruments are technologies directly affecting performance, training, or military decision making. Enabling instruments fill gaps in current technology and allow neuroscientific examination (laboratory) or evaluation (training or battlefield) of soldier performance, training, or military decision making. The committee feels this distinction is vital, because it is not immediately clear, for example, whether miniaturized signal processing technology will open additional opportunities to use laboratory devices currently considered impractical. If that happened, the miniaturization of signal processing would be an enabling technology.

All neuroscience technologies have spatial and temporal resolutions that define the neurophysiological building blocks they can study. Twenty-five years ago, the vast majority of detailed in situ function of localized structures in the human brain was extrapolated from work on animals measured with electronic stopwatches, clipboards, and scalp surface electrodes or was inferred from correlation studies of injury/pathology using psychiatric examinations. The introduction of noninvasive technologies has expanded the breadth and depth of studies of normal human brain function and allowed the development of noninvasive neural measurement techniques to study the functioning human brain. Figure 7-1 shows how these new technologies can monitor or even predict performance anywhere in the spatiotemporal plane. (For discussion of the history and advancement of neuroscience research related to Army applications, see Chapter 2.)

MANAGING THE SOLDIER'S PHYSICAL LOAD

There are multiple research and development opportunities involving soldiers, such as extracting information

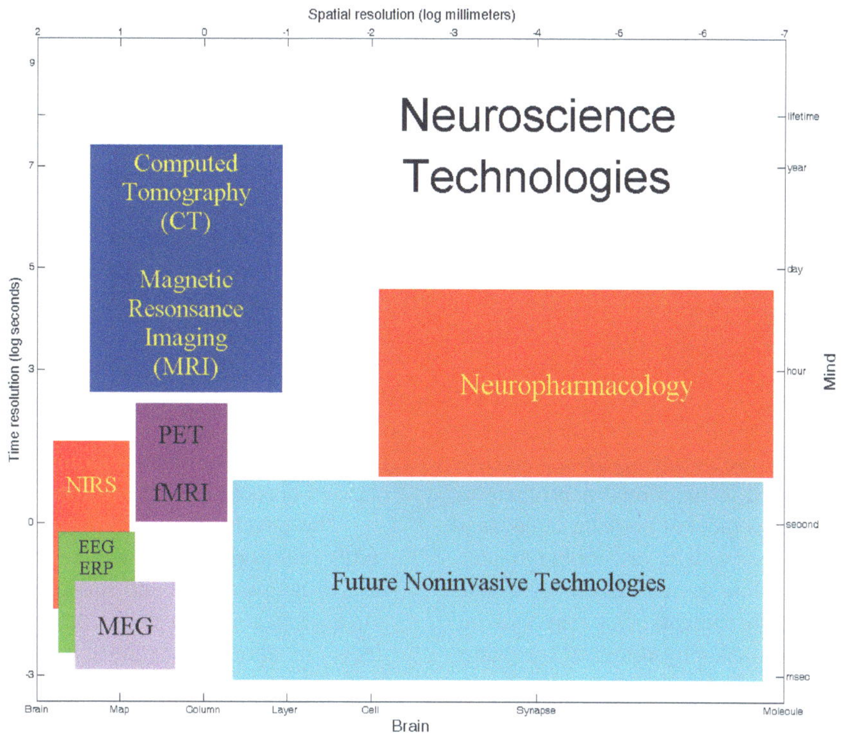

FIGURE 7-1 Various noninvasive imaging technologies provide insight into the brain (anatomy) and mind (function). The spatial resolution of a given technology defines the largest and smallest brain structures that can be observed, while the temporal resolution defines the elements of mind function to be measured. Academic and commercial research is primarily geared to improving resolution, although important measurements for the prediction of behavior can be made at any point in the brain-mind plane. Shown are several of the technologies discussed in Chapter 7. SOURCE: Adapted from Genik et al., 2005.

from the brain and nervous system, inferring neural states from physiological information, or designing control strategies to alter or enhance neural states. Nevertheless, the committee recognizes that critical ergonomic considerations limit the added burden—particularly added weight—that neuroscience technologies can place on an already overloaded soldier. Mission-enabling technologies (including devices for sensing, power, and onboard computating) must be considered as part of the larger system of a dismounted soldier's equipment load, and they should not add appreciable weight or volume to the helmet or backpack. A National Research Council study determined that any new device(s) should not add more than 1 kg to the helmet or 2 kg to the pack. More important, any helmet-mounted neuroscience technology should not interfere with ballistic protection, helmet stability, or freedom of head movement (NRC, 1997). The committee believes that these design and engineering constraints must be considered from the outset to ensure successful integration of a neuroscience technology with the soldier's existing equipment load.

MISSION-ENABLING TECHNOLOGIES

The Army has a basic requirement to process, distribute, and apply information efficiently. These requirements will only increase with the demands of a network-centric environment. Better cognitive performance must be achieved if soldiers are to contend with an ever-increasing river of information. Solutions are needed to address demonstrated operational requirements, such as avoidance of information overload and successful synthesis of information that selectively highlights the mission-critical features from multiple sources. The technologies described in this section apply knowledge and techniques from neuroscience to help solve these and related challenges in sustaining and improving soldier performance.

Mission-enabling (deployable) instruments or technologies of interest to the Army must be capable of being scientifically validated and include brain–machine interface (BMI) technologies, remote physiological monitoring to extend performance in combat, and optimization of sensor-shooter responses under cognitive stress. BMI technology examples include near-term extensions of current training applications of virtual reality (VR) systems, iconic or graphical information overlays on real-world visual displays, and various approaches to monitoring neurophysiological stress signals to manage information overload, and the use of neural signals to control external systems. Several of the technologies discussed may not appear to the casual observer to be rooted in neuroscience research. Where a connection is not obvious, it will be explicitly stated or the neuroscience aspect outlined.

Field-Deployable Biomarkers of Neural State

The first issue in applying laboratory neuroscience results to field operations is to find reliable indicators of neural state that can be used in the field (field-deployable biomarkers). The equipment used in functional neuroimaging laboratories is sensitive to movement of both the subject and metal objects near the subject as well as susceptible to interference from proximal electronic devices: Such constraints are antithetical to mission environments. One way to avoid this difficulty is to identify reliable physiological surrogates for the neural states of interest, surrogates that are easier to monitor in an operational environment. For example, alertness in a driving environment can be reliably determined by monitoring eyelid movement (Dinges et al., 1998). Neuronal state measurement is a primary topic in this chapter, but there are many other physiological indicators that can be evaluated for their reliability as biomarkers of functional brain states and behaviors. These include

- Galvanic skin response (GSR);
- Heartbeat, including rate and interbeat interval;
- Eye movements, including response times, gaze latency, and stability;
- Pupilometry;
- Low-channel-count electroencephalography (EEG);
- Cortical changes in blood flow, measured using near-infrared spectroscopy (NIRS);
- Blood oxygen saturation (also using NIRS); and
- Facial expression as monitored by optical computer recognition (OCR) techniques.

Combinations of these and future physiological measures are also possible. The committee believes that physiological indicators as surrogates for neurological states or conditions can be useful even before our understanding of the neurophysiological basis for the established correlation is complete. Therefore, the technology opportunities discussed here include the development and scientific validation of surrogate markers with research; to understand how they work is of secondary importance.

One futuristic technology that should be developed in parallel with work on individual physiologic indicators and surrogate markers is some kind of health and status monitoring tool for operational commanders that combines relevant neural measures to provide near-real-time feedback about operator neural readiness. To develop such a tool, a top-down functional analysis should be conducted to determine which of the neural indicators available are meaningful for different kinds of Army decision makers. For example, commanders on the battlefield could benefit from decision support that alerts them in near real time to issues with personnel neural readiness, such as unexpectedly high levels of fatigue or sleep-deprivation deficits in individuals or across units. Another class of decision makers that could benefit from readily accessible neural indexes would be medical commanders, who could decide how to allocate medical resources in response to rapidly changing events. Early versions of the monitoring system might find their initial application in training, including advanced training for specialized tasks as well as basic training of recruits.

The development of an operational neural health and status monitoring system represents an important intersection of neuroscience and military operational applications, since such a system could inform critical, time-pressured decisions about near-real-time soldier status. As the reliability and range of neural state indicators grows, this could also incorporate predictive biomarkers to aid commanders in comparing alternative scenarios. For example, a decision support tool that could indicate the neural impact of extending troop deployments, in both the near term and the far, could help to determine troop rotations and to select individuals or units for various activities.

EEG-Based Brain–Computer Interfaces

One area of neuroscience technology that has received much attention in the mainstream media is the development and use of EEG-based brain–computer interfaces. These interface systems have potential operational use in areas such as remote, real-time physiological monitoring—for example, a battlefield commander could receive some indication that a soldier is approaching maximum mental workload or stress. While such operational uses are possible, battlefield applications of these sensors as a neurotechnology are not likely to be realized in the next 10 years.

Commercial developers of EEG-based interfaces, which target primarily applications in video gaming and in marketing, generally claim that they can detect facial expressions, emotional states, and conscious intentions (Greene, 2007a). Their devices usually contain both the EEG sensors and machine-learning algorithms that require training for individual users. A handful of companies claim they will have brain–computer interface headsets commercially available in the very near future for gaming applications (Emotiv and NeuroSky are two such), and a similar number already claim success with methodologies for consumer research (in neuromarketing research applications) (EmSense, Lucid Systems, and NeuroFocus are three such methodologies). One technological advance that commercial EEG-based brain–computer interfaces have incorporated is the concept of a dry electroencephalograph, which is an EEG device that does not require the use of electrically conducting gel on the scalp. However, the capability of this technology has been questioned, since a dry EEG device cannot produce as strong a signal as a traditional gel-based electroencephalograph (Greene, 2007b).

Scientific proof of the claims for these brain–computer interfaces is virtually nonexistent, and they have been heavily criticized by academics (Nature, 2007; Greene, 2007b). The aforementioned companies have not published any scientific papers on their devices or methodologies, so the industry, as of now, is still extremely opaque. While the possible outcomes could be relevant for the operational Army, the use of EEG-based brain–computer interfaces to support real-time decision making should be considered basic research and funded accordingly. Furthermore, the headset technology demonstrated by companies such as Emotiv brings into question whether the primary signal that trains the interface is of cortical origin or based on cranial muscle polarization. These devices are therefore interesting as control interfaces to augment the number of devices a single soldier can control, but they probably do not qualify as a neuroscience technology opportunity.

One commercial application that could be more useful to the Army in the near term is a neurofeedback system for self-regulation. Neurofeedback, akin to biofeedback, allows individuals to receive visual and aural feedback about their brainwave activity to promote self-regulation. This technique, which was developed in part through research sponsored by the National Aeronautics and Space Administration (NASA), has been recommended as a therapeutic intervention for the treatment of attention-deficit hyperactivity disorder, traumatic brain injury, post-traumatic stress disorder, and depression in children and adolescents (Hirshberg et al., 2005).

While neurofeedback systems are not going to be used for operational real-time decision making anytime soon, their application in field settings could one day be of interest to the Army in ways that go beyond their obvious medical therapeutic benefits. CyberLearning Technology, LLC, which has an exclusive license with NASA for its neurofeedback methods, connects its system directly with off-the-shelf video game consoles such as the Sony PlayStation and the Microsoft Xbox. Given the ubiquity of these game consoles and personal computers in field settings, it may be possible to leverage this technology in the near term for use by soldiers in the field.

Haptic Feedback Technology for Virtual Reality

VR, a technology whose graphical base is driven by advances in the gaming industry, is now a common tool in behavioral neuroscience research and applications. Of particular importance for the Army is the use of VR for the study and modification of human behavior and for the enhancement of human abilities (Tarr and Warren, 2002). Indeed, VR is becoming a familiar technique for simulating flight, shipboard seamanship and navigation, and tank and armored vehicles. VR implementations are central to training for the Future Combat Systems program. The nascent Virtual Squad Training System is a wireless, wearable system that trains Army warfighters using simulated weapons like the Army Tactical Missile System in a virtual combat environment.

One area of VR that the Army has not yet exploited is the use of three-dimensional (3D) haptic interfaces. In general, a haptic interface provides cutaneous sensing and kinesthetic feedback to simulate the feel of physically manipulating what are in fact virtual devices. Haptic interfaces not only have training applications but also could be used for systems such as those that teleoperate robots.

The Army can leverage commercial-sector investments in haptic interfaces. The current commercially available haptic-sensing devices range from more extensive exoskeleton-based force-reflecting systems (e.g., Immersion's CyberForce) costing tens of thousands of dollars, to smaller, personal-computer-based, electromechanical force feedback systems (e.g., Novint's Falcon) that retail for a few hundred dollars. The larger force-reflecting systems could be useful in large simulation-training environments such as the Virtual Squad Training System. The PC-based systems, which are much smaller and less expensive, could be used for training during deployments.

Augmented Reality Technologies

Unlike VR, which seeks to replace the whole spectrum of sensory and thus perceptual experience with a simulated experience, augmented reality (AR) is a hybrid technology in which a display of the natural world is intermixed with information-rich virtual elements. A simple illustration of the basic approach is the display of information on, say, precipitation levels and wind intensity, derived from weather radar data, overlaid on a photograph of the terrain.

The principle of linkage is illustrated in the weather-map example by physical observation: Does the online image match ground conditions? The principle of scaling is illustrated by zooming in and out on the map: Time-locked weather patterns should stay constant over the same locations at all resolutions of an image. These weather depictions are fairly accurate, but it is still not unheard of for an online map to inaccurately describe what is observed with a quick trip to the window.

The first practical adaptation of AR for military use were "heads-up" displays in advanced fighter jets. These systems typically display on demand the status of the dynamics of the aircraft—including heading, altitude, and velocity—surrounding a central "pitch ladder" that illustrates the attitude of the aircraft itself. In most military aircraft displays, the augmentation includes assisted targeting capabilities.

Also, the display refers to the aircraft against a real-world background.

For dismounted soldier applications, an AR display might use a head-mounted, "monocle" device such as was deployed in a 120-person study of a simulated search-and-rescue operation (Goldiez et al., 2007). To conduct search-and-rescue operations in a building, team members must systematically clear an entire structure of possibly wounded compatriots and supply treatment if needed, while defending against or removing possible threats. Such operations tax working memory, and substantial improvement can be realized if mission elements requiring short-term memory encoding can be off-loaded to mobile technology. In the study cited, AR was employed to map a simulated building, which allowed participants to concentrate instead on tests of working memory such as locating mission-objective targets and planning for speedy exits. Similar applications subjected to more complex field testing will allow (1) smaller teams to complete the same missions, (2) teams to operate longer by shortening the time during which vigilance must be sustained, (3) team members to share graphic information in real time, and (4) teams to succeed in their mission even if they are operating at less than optimal cognitive capacity, perhaps as a result of fatigue.

Recent terrestrial applications of AR have focus-linked displays in which the augmentation (the information overlay) is spatially and temporally locked to some (usually geographical) aspect on the focal plane of the ambient display. These technologies have typically been used for way-finding and similar forms of orientation assistance. Such applications must be able to adjust scale as the real-world display zooms in and out and to lock on a real-world feature such that the display is constantly oriented correctly to a change in field of view.

AR poses some interesting neuroscience questions. An important concern is the correspondence issue: How is the electronically generated information to be permanently locked and scaled to the changing environment in which the AR user finds himself or herself? What happens when a failure of linkage produces a mismatch of either scale or orientation? Spatial disorientation is a problem that has been explored extensively in aerospace human factors studies. However, the latencies involved in spatial transformations and rotations can lead to a condition often called "simulator sickness" (Shepard and Metzler, 1971; Shepard and Cooper, 1982). Whereas the barrier to further advances in VR is largely insufficient computational capacity to generate the virtual world for the user, the comparable barrier to advances in AR is the difficulty of integrating information and of achieving perceptual realism in the display. Large investments will be required to overcome these more complex issues, but AR will eventually prove to be a more powerful technology than VR for incorporating neuroscience advances.

Additional sensory inputs will certainly be developed one day. Present technology for the senses other than vision is relatively rudimentary. A haptic simulation could include information overlay on a trigger that provides cutaneous feedback when a weapon locks on a potential target and has identified it as friend or foe. A soldier can be trained to set reflexive neurons in a state to pull the trigger without higher neural involvement, decreasing the time between acquiring and engaging the target. Such reflexive neuron training is in use today for complex, force-feedback motions like the double-tap.[1] Commercial AR applications include displays with visual, aural, or haptic channels that can be used individually or integrated. The type of heads-up visual display used in military aviation is being adopted in commercial industries such as trucking. The Army currently has a helmet-mounted monocular display as part of its Land Warrior program. However, the program has received mixed reviews (Lowe, 2008). It is possible that a variation on commercial heads-up display technology might improve the quality and performance of current implementations.

Commercial versions of visual AR include the Apple iPhone/iTouch system, which offers more intuitive and fluid gesture-based interactions. These same gesture-based interactions can be seen in Microsoft Surface, a tabletop interactive display that has recently become commercially available and was notably employed in television coverage of the 2008 presidential election. As an Army application, tabletop displays are primarily suitable for high-level, stationary command posts owing to their relatively high cost and fragility. Combined with appropriate software such as map search, visual AR devices in both small field-deployable versions and larger stationary versions enhance situational awareness very effectively.

Significant research has been done on enhancing decision making through use of an aural AR channel. Many human factors studies have stressed the importance of properly applying and integrating these aural systems. Innovative signal-presentation approaches for aural AR include spatial audio (reproducing spatial depth relative to the listener); ambient audio (either suppressing ambient noise or presenting local, external audio signals for operators in collaborative environments who are required to wear headsets); and continuous audio (mapping an alert condition to synthetically produced continual signals such as virtual rumble strips). Potentially useful Army applications include adaptation of commercially available spatial audio headsets and speaker systems that broadcast spatial audio in a group setting. The latter could be useful in enclosed environments such as a command and control vehicle (C2V). However, significant development work on the software would be required to adapt the hardware for Army use.

Visteon has produced the only commercial proximity- and touch-sensitive haptic controls. In these displays, which

[1]The double-tap is a combat pistol technique whereby the shooter sends a signal to the peripheral nervous system to pull the trigger on a semiautomatic pistol a second time once the trigger returns to a firing position. The second trigger pull is actually occurring while the shooter is locating the next target and assessing outcome using peripheral vision.

have been used in automobiles, as the operator's hand approaches the display, it lights up and a software-activated button provides haptic feedback to the user's touch, mimicking the feel of a mechanical button being pushed, even though the display is a flat screen. Potential uses for this technology include the C2V and control stations for unmanned aerial vehicles and unmanned ground vehicles.

Information Workload Management via Physiological and Neural Feedback (Including Augmented Cognition)

As noted earlier, neuroscience can help the soldier avoid information overload while helping with the cognitive tasks of synthesizing information and picking out mission-critical features. Examples of the latter include intelligence fusion and other forms of data interpretation to heighten situational awareness. The use of such information processing with presentation technology can enhance warfighter performance. Previous work by the military on this subject has dealt narrowly with filter methods and technologies to rapidly present the results to the soldier.

The Defense Advanced Research Projects Agency (DARPA) Augmented Cognition program (the AugCog program), which formally ended in FY 2006, sought to augment human information-processing capabilities through the design of interfaces incorporating neuroscience technologies that enable the interface to adapt in real time to the stress state of the user. Similar DARPA research continues under the rubric of Improving Warfighter Information Intake Under Stress. Army research along the AugCog path is continuing at the Natick Soldier Research, Development and Engineering Center. Appendix D reviews the phases of development work and testing under the AugCog program and the direction taken by Army follow-on activities. Highlights of the AugCog effort are presented here to illustrate the approach taken and the implementation achieved to date. The term "information workload management" refers to managing the presentation of information to sustain and enhance cognitive processing capability when the emotional-cognitive evidence indicates an individual may be reaching an overload condition. This information monitoring may feed back into an adaptive interface, as in the AugCog concepts, or, less ambitiously, it may trigger some type of warning signal to the user—for example, as part of an AR display.

The original goal of the AugCog program—to enhance information workload management through technologies that monitor indicators of cognitive state—is even more relevant now as the Department of Defense (DOD) moves toward network-centric warfare. However, while the researchers involved with the AugCog milestones made progress in terms of hardware and software advances, their results were preliminary, as one would expect. More important, perhaps, is the lesson that the original objectives of that program are not achievable in the near term because of barriers that became evident during this ambitious but early technology development effort.

Despite the stated goal of those closely associated with the AugCog program—that its technologies would be operational within 10 years—the likely horizon for an initial operating capability is much farther away. One major hurdle is development of a wireless EEG device that is unobtrusive, does not require the use of conducting gel (known as "dry EEG"), and is able to process onboard signals, all while the user is in motion and often under difficult environmental conditions, including electromagnetic interference. While some advances have been made in wireless EEG and dry EEG (see an earlier subsection on EEG-based brain–computer interfaces), the signals from these devices are substantially weaker than signals from more traditional electroencephalographs. Moreover, their ability to detect cognitive states for use in predictive algorithms in dynamic, uncertain environments has yet to be demonstrated and validated to the level required of an operational system.

The committee believes the Army should continue funding research in information workload management with a focus on hardware developments, including development of surrogate indicators for laboratory-based indicators of neural state, ruggedization of instruments for use in field environments, and advancement of associated signal-processing efforts. Without advances in these areas, the laudable information workload management techniques of AugCog cannot be operationalized. Substantial research and development will also be needed on predictive algorithms in dynamic, highly uncertain domains for open-loop systems with noisy sensor data. This is an instance where the higher-level technology used to monitor for and ameliorate cognitive overload will depend on the successful understanding of field-deployable indicators of neural state, as discussed above.

Technologies to Optimize Sensor-Shooter Latency and Target Discrimination

Another important Army concept applicable to a range of tactical combat situations is known as the sensor-shooter paradigm. The latency in sensor-to-shooter responsiveness is measured by the time needed to recognize a specific threat from its first appearance, to select the appropriate course of action to neutralize that threat, and to respond with the correct action. All other factors being equal, the lower latency from sensor to shooter will increase the efficiency of the sensor-shooter response. An equally important (or in some circumstances, more important) measure of response efficiency is target discrimination, which requires correct recognition of the threat with the fewest possible false negatives or false positives. The sensor-shooter should neither fail to recognize a threat (the foe) nor mistake a nonthreat (a friend or neutral actor) for a threat. Thus, improving sensor-shooter response efficiency requires optimizing the combination of a short latency period with a very high degree of target discrimination.

Complicating sensor-shooter efficiency is that one should not aim to minimize latency in a tactical vacuum devoid of appropriate strategic considerations. Strategy often involves longer-term goals for which a faster response (slower latency) is not necessarily better (Scales, 2006). The threat analysis technology that needs to interface with the soldier operationally requires strategic as well as tactical input and the ability to communicate both sets of information to the soldier so that he can make a decision.

Furthermore, the real-world scenarios to which the sensor-shooter paradigm applies often subject the soldier to stresses such as fatigue, sleep deprivation, information overload, and cognitive overload (Harris et al., 2005). Just the addition of a simple secondary cognitive task to be performed during a complex primary task will degrade performance of the primary task by an already overloaded individual from his or her baseline (Hancock and Warm, 1989). Technologies informed by neuroscience can boost the individual soldier's performance in sensor-shooter activities. The committee focused on two ways to support the sensor-shooter in difficult circumstances: devices to augment threat assessment and virtual simulation technologies to enhance intuitive decision making. The third aspect of the sensor-shooter paradigm, motor execution of the action decided upon, is an important research opportunity for the Army. It is discussed in Chapter 8.

Threat Assessment Augmentation Aids

It is clear that technology can be brought to bear on threat recognition. A number of devices are already used to support this function. Mainly they confer visual enhancement expressed on distal screens, on head-mounted displays that present a totally synthetic vision of the kind seen in VR, or on hybrid displays of the AR kind, including real-scale or telescopic capability, and overlaying relevant information from sources spanning the electromagnetic spectrum. These integrated displays attempt to replicate the perceived environment in some fashion. More sophisticated augmentation techniques can begin to replicate the capacity of the visual system of the eye and brain to focus on specific characteristics such as novelty, intensity, and context-driven importance. Such smart augmentation aids can help the observer focus on critical information.

As an example, consider the problem of detecting a recently emplaced improvised explosive device (IED) along a transportation corridor. One of the primary signals of threat is a change in the morphology or presence of roadside objects. Modern technology is very efficient at detecting a change in the scene if the only object in a field that changes shape or appearance is an IED. If the roadside litter in a field of view has also changed during the same time interval, however, detection may be markedly degraded. In the best case, a smart display could alert a patrolling soldier of change in the coming roadway scene, which could signal the presence of an IED. Such an augmented display could be programmed to scale these threats, e.g., pedestrian versus large-scale objects, to allow for a degree of preprocessed threat assessment. Biometric technologies capable of identifying specific individuals from a distance while supporting persistent visual monitoring of the environment will extend the amount of time available for a soldier to integrate fused data rather than collect and sort information, thereby increasing processing efficiency and reducing the likelihood of error.

Simulation Technologies to Enhance Intuitive Decision-Making Skill

Chapter 4 discussed decision making as it applies (primarily) to command-level decisions. Decision-making theory also provides useful insights into how simulation technologies can be used to help the sensor-shooter through the concept of intuitive decision making. The concept assumes that the decision maker has a high level of situational awareness. The simulations described in Chapter 5 that enable military leaders to accumulate life experiences that improve their intuitive decision-making skills can also be used to develop sensor-shooter training. Such simulations could be designed to adapt and respond to soldiers in an intelligent manner and portray cognitively, culturally, and intellectually accurate and challenging scenarios that identify, develop, improve, and assess these skills. Increasingly, human factors—the cognitive, cultural, and intellectual aspects of human conflict—are the main determinants of success on the battlefield (Scales, 2006).

A soldier–simulator interface that elicits personal interaction could lead to a self-referent memory approach by the trainee, increasing the accuracy of an individual's recall when a similar situation is faced again (Rogers et al., 1977). This type of interaction with the simulator would be based on the theory of recognition-primed decision making (Klein, 1989), and if the interaction is properly exploited with well-designed interfaces will lead to perceptual learning in the areas of attentional weighting, stimulus imprinting, differentiation, and unitization (Goldstone, 1998). A well-designed decision-making simulator could enable soldiers who will have to function in demanding sensor-shooter roles to learn using scenarios that provide life experiences, bloodlessly.

Finally, as discussed in Chapter 4, recent advances in neuroimaging enable researchers to follow the spatial pathways and temporal patterns of expert decision makers. For example, the detection of potential threats as revealed in VR displays appears to involve the amygdala and related brain regions. While most of the new information and correlations have been achieved in the laboratory environment, new lightweight, portable technologies that take the place of functional magnetic resonance imaging (fMRI) and magnetoencephalography (MEG) to detect loss of decision-making capability would enhance operational skills and survival on the battlefield. These new technologies are expected to be of greatest benefit in High OpTempo[2] environments (continuous operation for of 12-36 hours). This is another example where research and development work on field-deployable sensors to indicate neural state is essential for achieving a more advanced state of neuroscience-based technology.

RESEARCH-ENABLING TECHNOLOGIES

Several of the mission-enabling technologies in the preceding section require for their further development a fuller understanding of common neurophysiological patterns in human behavior. Research-enabling technologies are also needed to develop tools to study and assess underlying aspects of performance such as ground-truth workload and attention to detail. The advances made with research-enabling technologies will be deployed with soldiers on a limited basis, or used in training, simulations, or laboratory environments. Some of the mission-enabling technology described in the previous section will also find uses in the research environment.

Investment in research-enabling technology is crucial for adapting current technology to Army applications, as well as for advancing to future generations of Army applications. Such technology might also help the Army conduct scientifically rigorous validation and testing for emerging mission-enabling technology. Some of the opportunities simply involve bridging gaps in technology—for instance, the ability to use fMRI and concomitant eye tracking across a wide visual angle. In this section the committee discusses signal processing challenges, control strategies for BMIs, fatigue and sleep models for soldiers, advances in functional paradigm technology, adapting laboratory neuroimaging technologies for use in the field, and data fusion. The committee also touches briefly on the science of connectonomics. It looks at the development of a few pieces of hardware and imaging methodologies that could dramatically advance several basic science techniques.

Signal Processing Challenges

At present the methods for extracting information from the brain and nervous system can be divided into two categories: invasive and noninvasive. Invasive methods include multielectrode recordings, local field potentials (lfp's), and calcium imaging. The advantage of the invasive methods is that they provide the most direct information about the functioning of specific brain regions on a very fast timescale. The disadvantage is that they frequently require surgery and sometimes cannot be used in humans. Noninvasive methods include EEG, MEG, diffuse optical tomography (DOT), diffusion tensor imaging (DTI), fMRI, GSR, electromyography, and electrooculography.

Noninvasive recording techniques have the advantage of not requiring an invasive procedure to place the recording apparatus. However, they frequently require a tremendous amount of additional hardware and infrastructure to collect the information. In addition, noninvasive procedures generally allow high resolution on the temporal scale at the cost of less resolution on the spatial scale or vice versa. The disadvantage of noninvasive recording techniques is that the information they collect is often indirect and less specific.

The first conceptual issue surrounding the processing of signals is our limited understanding, for each of the invasive modalities, of what information the signals are providing about neural activity in a specific brain region and the relation between that activity and specific physiological changes and/or behaviors. Addressing this issue requires executing specific experiments and developing specific techniques. Research in neuroscience has not completely answered the challenging signal processing questions that must be answered if the data from invasive monitoring modalities are to be used efficiently. Among these questions are the following: To which aspects of a stimulus does a neuron respond? How do groups of neurons represent information about a biological signal (a movement command, a memory, a sound or light input) in their ensemble spiking activity? How can the plasticity in single neurons and ensemble representations of information be tracked reliably across time? How should algorithms be devised to process the activity of large numbers of neurons in real time? What sort of signal processing and biophysical information should be used to optimally fuse information from different types of recording techniques?

The second conceptual issue is the extent to which brain activity from invasive measurements can be related to brain activity inferred from noninvasive measurements. If the relationship is strong, the noninvasive technique might be an adequate stand-in for the invasive technique and could lead to application as a field-deployable surrogate biomarker. If the relationship is weak, certain types of brain information may not be accessible by noninvasive means. These observations point to the need for simultaneously conducting invasive and noninvasive recordings in order to understand the relation between the two.

For example, EEG is the simplest and perhaps the most widely used noninvasive neural recording technology. Although EEG has been used for nearly 80 years to study brain function dynamically, how it works is not completely understood. Much of the use of EEG signals still depends on heuristic associations. The fundamental questions here

[2]High operations tempo (OpTempo) refers to missions carried out as quickly and fully as feasible, to apply overwhelming force in a time frame such that opposing forces are unable to respond and counter effectively. By their nature, High OpTempo missions are characterized by high levels of psychological and physical stress, including constant awareness of mortal danger and potential for mission failure, combined with heavy decision-making loads. When High OpTempo is combined with sustained operations (SUSOPS) (missions lasting longer than 12 hours before resupply), cognitive capabilities are easily overtaxed and prone to degradation or failure.

are, What does an electroencephalogram mean? What is the biophysical mechanism underlying its generation? To what extent can it give us reliable information about both neocortical and subcortical activity? Studies that combine EEG and invasive electrophysiological recordings in specific brain regions will be required to answer these questions. MEG is used less often than EEG, but similar questions can be asked about it.

In the last decade and a half, fMRI has become the fundamental tool in many fields of neuroscience. The basic question—How do changes in neural activity relate to the changes in local blood flow and local blood volume that are necessary to produce the fMRI image?—is only beginning to be answered (Schummers et al., 2008). Similar fundamental biophysics questions about DOT have yet to be answered. In addition, when it is possible to combine a high-temporal-resolution technique such as EEG with a high-spatial-resolution technique such as fMRI, what is the optimal strategy for combining the information they generate? This example illustrates that simultaneously recording using two or more noninvasive methods can also be mutually informative.

The third conceptual issue is that the ability to analyze behavior and performance quantitatively is essential to understanding the role of the brain and the nervous system in their guiding function. Some typical measures of performance include reaction time, GSR, heartbeat dynamics, local neurochemistry, and quantitative/objective measures of pain and nociception.[3]

In most behavioral neuroscience investigations performance is measured along with neural activity using one of the invasive or noninvasive methods. These investigations are crucial for linking neural activity in specific brain regions with overtly observable measures of performance and physiological state. Often the analyses of these performance measures are quite superficial and not very quantitative. For example, reaction times are simply plotted rather than analyzed with formal statistical methods. Similarly, GSR and heartbeat dynamics are directly observable measures of the brain's autonomic control system. Such signals are rarely if ever analyzed as such. fMRI studies are beginning to help us better understand the processing of pain and the signals from the body's pain receptors (nocioceptor signals). For this work to translate into techniques that can be used to aid the military, quantitative measures of pain and nociceptor stimuli must be developed.

The fourth signal processing issue is being able to properly fuse information from different sources, whether invasive or noninvasive, and the fifth issue surrounding signal processing is the challenge of rapid and (ideally) real-time visualization and analysis. In short, the ability to effectively use information collected from the brain and nervous system to enhance performance and improve therapies depends critically on the signal-processing methods used to extract that information.

All of the popular noninvasive methods for measuring neural states in humans have unanswered questions concerning their underlying neurophysiology. Although one can certainly glean useful methodology without probing deeply, fundamental questions remain. If research answers them, more applications and measurement techniques may open up, including, eventually, field-deployable indicators of neural state.

Fatigue and Sleep Models for Soldiers

Chapter 5 talks in detail about fatigue and sleep research, as well as mitigation strategies. Two important technologies enabling the performance-sustaining research discussed in these areas are (1) the computational models for predicting behavior and (2) the physical models for transferring results to the appropriate warfighter population.

The computational model used is a vector of parameters important for sleep or fatigue, and inputs calibrated to a specific individual soldier. The additional strategies the officers employ in the field—naps, nutritional supplements, etc.—should be included in the model, and it should account for the difference between an academic research subject used to construct a model and a soldier.

Ideally, the physical model used in research would be an actual soldier in the state of readiness expected at the start of a mission. However, the multitude of research variables that must be tested necessitates using an ordinary civilian to stand in for the soldier. Chapter 3 described an opportunity to leverage research using high-performance athletes. In an academic setting, it would be preferable to use persons from a university community for most of the studies and reserve actual soldiers for experimental runs once the paradigms are well understood and being tested for validity.

Research in the area of fatigue might also include a systematic study of the differences between cognitive fatigue, physical fatigue, and fatigue, including environmental stress, from hypoxic or thermal challenges; biomarkers predictive of a soldier's susceptibility to fatigue under extreme environmental conditions; and behavioral measures of fatigue to advance screening and testing procedures for soldier assessment.

Functional MRI and Hardware to Support fMRI Research on Army Applications

Functional MRI is detailed in Chapter 2. Technology associated with fMRI for use in clinical health care is receiving sufficient investment from industry. However, clinical applications require only medium spatial resolution (3-4 mm) and low temporal resolution (tens of seconds). These resolutions are usually sufficient for a clinical deter-

[3]Nociception is the physiological perception of a potentially injurious stimulus.

mination of whether a major circuit in the brain is functioning normally; however, they are inadequate for measuring neural responses to instantaneous events in rapid succession, among other research paradigms.[4] Academic research laboratories, funded mainly by the National Institutes of Health, possess fMRI technology that is superior to the equipment available commercially (~2-mm and 1-sec resolutions for whole human head scans). This improved spatiotemporal resolution is primarily achieved through use of advanced imaging electronics, such as parallel signal receiver channels rather than an exclusive concentration on ever-increasing static field strength. Cutting-edge laboratories have advanced measurement techniques that are a vast improvement over conventional imaging, but even typical facilities have invested in excess of $10 million for equipment and facilities, an investment that could be leveraged by the Army for the evolutionary application of current technology. The Army needs to monitor advances in existing facilities and consider ways to utilize them.

Some areas of research could be of great value for Army applications but are not being addressed by industry or academia because they have little if any potential for use in the clinical market. These areas are likely to require Army investment to achieve sufficient understanding to adapt results from laboratory environments to the field. They include vertical-bore MRI; full-motion, interactive stimulation; wide-angle, immersive visual stimulation; and high-temporal-precision stimulation and monitoring.

Currently, all fMRI research is done with the subject lying down. There are physiological and perceptual differences between horizontal and vertical orientation. To determine whether and to what extent supine-orientation fMRI is applicable to field situations, the least that must be done is to conduct experiments with the subject sitting up or, possibly, standing. Subjecting participants to heat, humidity, smells, and other such stimulants encountered in combat situations will also be required. This necessary work will require designing and building a specialized MRI machine, with its supporting laboratory, that is capable of scanning subjects in the vertical position while also exposing them to relatively rapid environmental changes.

Developing such an fMRI system is likely to entail an investment horizon of at least 5-10 years. One company (Fonar) produces vertical MRI machines for humans, but these machines are primarily for orthopedic imaging rather than brain imaging and lack the high temporal resolution needed for Army-relevant research. At least one Army application laboratory would be required.

The committee estimates that setting up a facility to perform vertical fMRI at 3 T or more using state-of-the-art imaging systems would cost $10 million to $20 million for the first 5 years of operation and, nominally, $2 million per year thereafter. Collaboration with external partners could reduce the Army investment. If the first such machine proves useful for Army applications, additional machines should cost substantially less.

The main risk in this investment is that the study results may show there is no additional information to be gleaned for Army applications by examining subjects sitting or standing rather than lying supine. Although such a result cannot be ruled out before the requisite testing is done, it would contradict current research on perceptual differences observed in nonhuman primates. Mitigating this risk is that measurements are not expected to be any worse than with a commercial off-the-shelf[5] system; however, the custom system is not expected to be a versatile clinical machine.

Moreover, the majority of research paradigms in fMRI are static, meaning that stimulation is planned out entirely before the experiment. A small number of laboratories have produced technology that allows for feedback based on subject responses or physiological reactions that helps determine the subject stimulation in real time. This real-time technology would allow basic research into neural function in more naturalistic environments. Pioneering research in this field is being carried out at the Army's Institute for Creative Technologies.[6] The goal here is to continue such research and offer more naturalistic environments for research paradigms. A real-time system should be able to log the time at which a stimulation occurred or a response was made, including eye movements, with an accuracy of less than 1 msec and a precision of 500 μsec. This software environment should be deployable to neuroimaging centers doing Army research, requiring relatively small amounts of hardware ($200,000 for research quality and of $50,000 for clinical quality) and local technical support. The committee notes that this advance in real-time, interactive paradigms for neuroscience research should be developed with a vertical fMRI capability but would also be applicable to the development of standard supine-oriented machines.

The Army should also support development of extended-range visual stimulation hardware that is closer to combat conditions and also MRI-compatible. The hardware currently available for fMRI includes wide-angle immersive visual stimulation and high-frequency presentation, but the current state of the art is ±15 degrees of nominal center (Cambridge Research is using ±45 degrees but has not yet publicly demonstrated the capability), and peripheral stimulation standards normally exceed ±40 degrees of center.

[4]This information comes from two installed software packages (BrainWave, GE; and Neuro 3D, Siemens) and from discussions with company representatives attending the annual meeting of the Radiological Society of North America about what is expected to be released in the next few years.

[5]Examples of such systems would be research scanners made by the three main commercial vendors: Siemens, General Electric, or Philips.

[6]Jonathan Gratch, Institute for Creative Technologies, University of Southern California, "The Neuroscience of Virtual Humans," presentation to the committee on February 13, 2008.

Additionally, 60 Hz is the standard display rate for research, with some optional displays claiming 100 Hz.[7] The video gaming industry's top-of-the-line displays are five times faster (500 Hz). It is unlikely that 500-Hz displays would be required for fMRI research in the next 5 or 10 years as long as the screen refresh time is known to the submillisecond standard.

Finally, eye-monitoring hardware that tracks gaze also mainly exists in the 60-Hz world. In order to correct reaction times for eye movements, cameras sampling at 1000 Hz and above will perform the best. State-of-the-art hardware for investigating behavior provides 1250-Hz sampling rates; however, these cameras are not intrinsically MRI-compatible. The high-speed solution is to utilize a limbus tracker,[8] which can sample as high as 10 kHz but has neither the angular resolution of a high-speed camera nor complete pupilometry capability. Additionally, even high-sampling-rate equipment has latency due to its USB PC interface, giving an overall synchronization uncertainty of 8 msec (ideally it should be negligible). The various components of these eye-tracking systems exist at separate facilities, but engineering them all into a standardized setup for Army applications research would be a worthy investment.

Transferring Laboratory Neuroimaging Technologies to Field Applications in the Far Term

The gold standard in functional neuroimaging is currently fMRI. Details of this technology are introduced in Chapter 2. To summarize: fMRI indirectly measures neuronal activity by watching changes in local blood flow around active neurons through the blood oxygen level-dependent (BOLD) effect. BOLD changes can also be observed by NIRS—also known as diffuse optical tomography (DOT)—detectors, though at a lower resolution. The underlying neuronal activation may also be observed at low resolution noninvasively with EEG or MEG. The committee expects that field-deployable fMRI technology will not be available for at least 20 years. Accordingly, results from the high-resolution fMRI laboratory experiments will need to be translated to field monitoring applications through the use of surrogate markers well-correlated with the fMRI results. The most likely candidates for field deployment are NIRS/DOT, EEG, and transcranial magnetic stimulation (TMS).

There are a number of approaches to portable, field-usable application of fMRI-based neuroscience research:

- Direct measurement of BOLD responses in the brain,
- Direct measurement of the neuronal firing that caused the BOLD response,
- Suppression of unwanted brain activation, and
- Enhancement of desired brain activation.

The committee projects that none of these approaches will be practical before the far term at the soonest. However, breakthroughs happen, and all of them could have great impacts in the future and deserve to be monitored or even considered for some initial pilot funding.

BOLD responses could be measured directly with in-helmet NIRS/DOT detectors. NIRS/DOT can detect the dynamic changes of any spectroscopically active molecule. Single- or dual-wavelength techniques are usually employed to track blood flow in the brain, providing a crude monitor of the BOLD effect. Further development of these techniques over the next 10-20 years will lead to portable systems to take advantage of results of basic brain research in the intervening years. Measuring neuronal firing in the field is a long-term goal no matter if it will be achieved with a few sensors or with a several-hundred-channel electrical imaging system.

Quantitative EEG (qEEG) is a marketing term to describe the marriage of traditional EEG with the digital recording and analysis of signals. The nomenclature change is promoted mostly in legal circles to add weight to expert testimony in civil tort proceedings or criminal defenses ("my brain made me do it"), as well as alternative medicine circles that use biofeedback to treat physiological ailments. Most of the claims for qEEG (sometimes labeled rEEG) are suspect; however, there is solid science behind the decades of analyzing the signals detected by transient EEG, and this area of neuroscience research is well worth monitoring.

If research into the real-time processing of transient EEG ever reveals something of value, then a deployable in-helmet EEG detector would need to be available. Immediate uses for data on general sleep and fatigue are envisioned that would justify deploying existing EEG equipment long before a high-sampling-rate, 100-channel system is needed. A portable monitor was demonstrated at the 2008 annual meeting of the organization for Human Brain Mapping.[9] A field-deployable EEG detector system has in fact been developed and should be tested in the field. The 5-year goal is the recording of a half-dozen channels with the subject jogging on a treadmill, and the 10-year goal is producing a

---

[7]In liquid crystal display technology, unlike with the older cathode ray tube (CRT) displays, merely pumping up the input video frequency does not result in faster displays. The liquid crystal elements have a limit to their on-off transition time—typically 15-20 msec for a standard desktop. This transition time explains why flat-panel displays do not flicker like CRTs and therefore cause less eyestrain. Top-of-the-line gaming displays can transition in as little as 2 msec, providing a true 500-Hz refresh.

[8]A limbus tracker illuminates the eye with infrared light (IR) and uses a single photodiode to collect the IR reflection. The motion of the edges of pupil and iris induce changes in the total reflected IR intensity. Standard eye trackers use a camera to transmit IR video of the pupil, iris, and sclera, which is processed using image analysis software. Limbus trackers are good at detecting any motion of the eye but do not provide any directional or absolute gaze information.

[9]This was the 14th Annual Meeting of the Organization for Human Brain Mapping (Melbourne, Australia, June 15-19, 2008). The meeting information is documented at http://www.hbm2008.com.

system that can be used in real-time training and assessment exercises.

Unwanted brain signals can be temporarily suppressed by noninvasive means in the laboratory using TMS. TMS uses high-frequency magnetic fields to block the functioning of target neuronal structures, in essence jamming the functional ability of a brain region. Two aspects of this technology need to be worked on: targeting smaller areas to lessen the side effects and making the technology deployable in a vehicle or helmet. Additionally, much research is required to learn which brain signals should be blocked and under what circumstances. Finally, there has been little research on the long-term impact of multiple TMS exposures on brain circuitry, leaving significant ethical concerns about exposing healthy humans to this technology over long periods.

Enhancing desirable brain networks is usually accomplished with neuropharmacology, as discussed in Chapter 5. Additionally, it is possible that TMS can be employed to enhance rather than suppress activation. One recent study showed enhancement of top-down visuospatial attention using combined fMRI/TMS stimulation (Blankenburg et al., 2008). The ability to target smaller areas is an objective sought by the TMS research community in general, but making such a device deployable in the field would require Army investment. Making this technology available in-vehicle is achievable in the medium term. The committee believes that in-helmet TMS technology would not be a useful investment until definitive applications, enhancing or inhibiting, are identified in the laboratory.

Implantation of deep-brain stimulators has been researched for use in Parkinson's, epilepsy, and obsessive-compulsive disorder for both suppression and enhancement of neuronal activation. Study of such an invasive technology should be limited to the treatment of similar disorders in soldiers.

Finally, although it is unlikely that a portable fMRI for detecting BOLD can be developed in the next 20 years, a low-field, combined fMRI/MEG approach that would measure both direct neuronal currents and BOLD fluctuations could produce a soldier-wearable system. Initial laboratory experiments with fMRI/MEG (McDermott et al., 2004; Kraus et al., 2007) indicate some feasibility, although it will require substantial technology development and breakthroughs in both ultralow magnetic field detection and signal capture in electromagnetically noisy environments. The committee concluded that such developments are equally unlikely in the next 20 years. However, the fMRI/MEG approach should be monitored, as it is already being supported by the National Institutes of Health and the Department of Energy despite the risky prospects for the technology.

An interesting outgrowth of the low-field fMRI/MEG direct neuronal firing work is a high-field application of the same method using the parallel acquisition mode of an advanced brain imaging coil. The principle here is detection of stimulated magnetic resonance relaxation at very fast repetition times: up to 100 frames per second. This allows very high temporal resolution of BOLD signals and can calibrate individual BOLD characteristics. This technique, termed inverse magnetic resonance imaging, could be very valuable in understanding fundamental brain activity (Lin et al., 2008).

An emerging imaging technology known as DTI is an enabling technology for a new field known as connectomics, the study of the brain's neural pathways for information transfer. The name derives from the concept of the human connectome—the entire collection (body) of neural connections—in much the same way as the entire collection of genes is termed the human genome. (See Box 5-3 in Chapter 5.) Connectomics is an area of basic neuroscience research with tremendous potential to enable the understanding of brain function, and DTI may have potential for future Army research.

The Army should also monitor research on atomic magnetometers for its potential to contribute to portable and rugged MRI (Bourzac, 2008). Atomic magnetometers may prove of great importance to MEG, and MEG imaging is the basis for inverse MRI, which will need to be developed for ultraportable (less than 20 pounds) MRI scanners. However, putting 100,000 sensors around a soldier's head does not make much sense unless you can deal with all of the sensor information in real time. Although this technology cannot support Army applications until the signal processing issues outlined in a previous subsection have been addressed, the committee views the area as a future opportunity.

Optimal Control Strategies for Brain–Machine Interfaces

One far-term technology opportunity will require a great deal of technique development and experimentation—namely, the extension of current control theory and control technology to optimal strategies for controlling an external system through signal communication only (an information interface) between the brain and the external system's control input and feedback subsystems. The natural way our brains control an external system is through efferent peripheral connections to muscles, where the information signal is transduced through a motor response; for example, we turn a wheel, step on a pedal, press buttons, move a joystick, utter a vocal command, or type a command to a software subsystem on a keyboard. The external system provides feedback to the controller-brain in the form of sensory stimuli: visual information, proprioceptive inputs, auditory signals, etc. In a BMI, control signals from the brain are identified and transmitted by a decoding subsystem, which communicates the signal to the external system's control input interface. In addition to the customary range of feedback cues via peripheral sensory stimuli, the external system could in principle send feedback signals to the brain through stimulation channels.

In this sense of information transmission between the controller-brain and the controlled external system, a BMI can use either invasive or noninvasive technologies for control signal monitoring and (possibly) feedback stimulation. (See the discussion of invasive and noninvasive monitoring methods in the subsection on signal processing above.) Invasive control and feedback methods are most relevant to technological aids to recover normal function lost through an accident, disease, or combat, and should, for ethical reasons, remain restricted to such applications, which would include advanced prosthetic limbs and, perhaps, alternatives to a limb such as a directly controlled wheelchair or a reaching-grasping device. In the context of this report, however, the types of external systems to be controlled are not prostheses but the kinds of systems a soldier would normally control by efferent motor responses: a vehicle, a UAV or UGV, or an information-processing system/subsystem (i.e., a computer or a microprocessor-based information node). For such systems, noninvasive (as opposed to invasive) control and feedback methods are, for the foreseeable future, the only practical and ethical options.

The entire field of BMIs is at an early stage of understanding. For example, we are just beginning to learn about the incredible potential offered by the plasticity of even a mature adult brain. There is much that will need to be learned from the current and continuing work on invasive methods for prostheses before we can even think about the longer-term challenge of embedding BMIs.

Advanced Upper-Limb Prosthetics

With improvements in battlefield medicine, many more soldiers than in previous wars are now surviving serious injury. Many of these injuries involve loss of an upper limb or of multiple limbs. While leg prosthetics have been very successful, the prosthetics for an upper limb, which has over 20 degrees of freedom, are a much greater challenge. Current versions of upper limb prosthetics, which use electromyography activity to control the limb, have limited degrees of freedom, are difficult to control, and are very heavy, expensive, and uncomfortable. Indeed, patients often abandon these complex prosthetic limbs for simpler and more rudimentary limbs.

An exciting long-term goal for Army medical research is to develop upper-limb prosthetics that are neurally controlled. The two most promising approaches for the limb control (efferent) and sensory feedback (afferent) interface with the nervous system are connection to the peripheral nerves or directly to the cerebral cortex. The peripheral nerve approach involves recording signals from the stumps of the severed nerves to control the prosthetic limb. The cortical approach requires a direct BMI that records the signals derived from activity in the motor and sensory-motor areas of the cortex involved in forming movement intentions. Both approaches require not only "reading out" the movement intention (efferent signaling) of the subject but also a means of sensory feedback (afferent signaling), which is essential for dexterous control.

There are several challenges for technology development in designing brain–machine interfaces for upper-limb prosthetics using cortical control. One is to design implants that ensure longevity of recording, ideally for the lifetime of the individual. Current implants typically last only a year or so; however, some implants have lasted for a number of years. Understanding biocompatibility and durability and other factors that affect the longevity of implants is an important area of research. A second developmental challenge is the integration of electronics with electrodes. Ideally the substrate of the electrodes should be metal leads on a silicon substrate. This type of electrode can easily be integrated as a single unit with integrated circuit electronics.

A third challenge is to make the electrodes movable—that is, able to automatically search out cells for optimal recording and move to new cells when cells are lost. A fourth challenge is the use of local field potentials and spikes to improve recording decodes, particularly for determining behavioral states and transitions between them. Fifth, implantable electronics need to be developed that allow on-board processing and decoding of neural signals and wireless transmission of the signals to controllers within the prosthetic limb. Finally, limb robotic technologies need to be advanced to achieve lightweight limbs that can generate forces similar to those in natural limbs and with power sources that can last for long periods of time before recharging.

Given all of the challenges and the delicate nature of direct neural connection technology, it is unlikely that the interfaces could be made battlefield robust in the foreseeable future. Invasive technology is currently utilized in medical prosthetics, including direct brain connections such as multichannel cochlear implants to replace the sense of hearing. These direct connections are able to capture and generate individual neuronal currents, as well as monitor or induce coherent neural activity at much greater signal-to-noise ratio than noninvasive technology. The invasive technologies should therefore be considered the best possible case of both signal detection and interface complexity for what could be achieved via noninvasive technologies. These medical applications are therefore important for the Army to monitor as a guide for what may be possible noninvasively.

Other Prosthetics Applications with Relevance to Brain–Machine Interfaces

In addition to BMI systems to facilitate recovery of motor function, other prosthetic devices are on the horizon. These include devices that carry out deep brain stimulation to improve cognitive function and arousal state and to treat depression. There has also been work recently on central auditory neural prostheses that stimulate the inferior colliculus and visual neural prostheses using stimulation in the retina or lateral geniculate nucleus. In most of these cases, research has demonstrated the feasibility of devices either stimulating a given brain region or using information from a particular brain region. A fundamental question must be answered: What are the optimal control strategies that allow a prosthetic device to interface in the most efficient and physiologically sound way with its human user? As a simple illustration, most deep-brain stimulation to treat Parkinson's disease is carried out by applying a current once the stimulator is implanted. Given what we know about neural responses and, in particular, about neurons in the subthalamic nucleus, is it possible to design a device for stimulating this brain region that does not require the constant input of current?

SCIENTIFIC AND TECHNICAL BARRIERS TO NEUROSCIENCE TECHNOLOGIES

Chapter 2 discussed ethical and legal barriers to neuroscience research and development. There are also scientific and technical barriers to the development of neuroscience technologies that could be overcome using advances in unrelated fields of science and engineering. Advances in the miniaturization of electronics and other components, for example, would enable development and deployment of research-enabling imaging technologies needed to substantiate and apply neuroscience hypotheses in the field. Such advances would also facilitate the design of less bulky and ungainly BMIs. Add biocompatibility to bulkiness as another barrier to the development of neural prostheses. Once this barrier is overcome, biocompatible devices could serve as alternatives to more invasive monitoring and imaging techniques.

Data fusion is yet another barrier. Neuroimaging data collected by various means will not realize their maximum utility until different modalities can be fused. Additionally, even though one may be able to fuse laboratory results, field-deployed equipment will have its own measurement quirks that must be taken into account when the task of fusing data is transferred from the laboratory to the field.

Possibly the greatest challenge for the Army is to ensure that its institutional expertise—in, for example, analysis modalities and data fusion techniques—resides in individuals of all ages. Overcoming this barrier should be a major goal for the Army. The committee observed that much of the neuroscience expertise in the Army is possessed by late-career scientists without mid- and early-career backup. This failure to diversify agewise puts the Army at the risk of losing substantial institutional intellectual equity each time a senior neuroscientist retires. In-house expertise is crucial for leading research in Army-specific areas, such as understanding the amount of effort involved in the measurement of ground-truth,[10] knowing whether it is possible to train up to an arbitrary capacity (versus improving a human–machine interface, for example), and recognizing further technology opportunities.

[10]Ground-truth workload is an objective measure of brain activity based on functional neuroimaging or a corresponding field-deployable biomarker. The phrase "ground truth" relates to the fact that most measures of workload are based on subjective response to questions such as "On a scale of 1 to 10, how busy did you feel?" This objective measure is a goal for the future.

TRENDS IN NEUROSCIENCE TECHNOLOGY

The committee identified several trends in neuroscience technology that the Army should monitor for application to its needs. Advances in neuroscience technology and methodology are occurring at an extraordinary rate, and extraordinary measures are needed to keep abreast of developments in the field. The committee identified trends in six areas: cognitive psychology and functional imaging, targeted delivery of neuropharmacological agents for operational—that is, not medically indicated—purposes, multimodal fusion of neural imagery and physiological data, new types of averaging in fMRI, database aggregation and translation for meta-analyses, and default mode networks.

Cognitive Psychology and Functional Imaging

Because fMRI has become so widely available, psychologists are able to test cognitive models of the human mind against functional data. Traditional cognitive psychology developed and flourished well before scientists could noninvasively image activity in the human brain, and some of them still deny the utility of knowing which areas of the brain are active at particular times. They use the analogy that knowing which parts of a computer are consuming the most energy does not tell you what the software is doing.

Be that as it may, fMRI continues to produce consistent results on psychological fronts that cannot be dismissed, and the imaging community is more and more accepting the need for theoretical models of cognition (Owen et al., 2002; Heuttel et al., 2004) such that the amount of functional data available is overwhelming resources that could otherwise be used for experimentation. The Army should monitor the collaborative progress that is made among neuroscientists for synergies that may reveal possible future opportunities for applications.

Targeted Delivery of Neuropharmacological Agents for Operational Purposes

An important technology trend is improvement in the ability to deliver pharmacologic agents to specific brain locations in the nervous system in a controlled manner. It is hypothesized that targeted delivery mechanisms will open up significant new classes of compounds for use above and beyond those considered safe for oral ingestion. Research in this area is of two kinds: (1) the identification of specific functional targets in the brain and spinal cord and (2) the creation of delivery systems that can place pharmacologic agents at these targets. Besides ingestion, there are three routes by which drugs can be delivered to the nervous system: by injection, inhalation, and topical application. With all of these routes it is important to know whether the objective is to enter tissue, which is an aggregation of cells, or to enter cells directly. For any intravenously delivered agent, a key issue is passing the blood-brain barrier. Several new modes of drug delivery are now being studied, including encapsulation in nanoparticles and scaffolding in polymer systems (Lee et al., 2007; Cheng et al., 2008). Site-specific delivery can now be controlled more precisely by targeted activation and inactivation. It is apparent that this is a very active area of research that will see many improvements over the next several years. Delivery systems are key technologies that the Army should monitor rather than invest in directly.

Multimodal Fusion of Neural Imagery and Physiological Data

Another trend is the collection of data from several physiological monitors concurrently or the fusing of data from separate sources into a common paradigm. An important trend in many imaging centers is the development of functional neuroimaging tools to fuse multimodal images using various combinations of fMRI, DOT, EEG, and MEG measurements. An EEG-based instrument has recently been commercialized to allow data from MEG and EEG or fMRI and EEG to be collected simultaneously. These imaging and electrophysiological measurements can also be combined with other physiological variables such as heart rate, GSR, eye movements, blood pressure, and oxygen saturation. Other instruments combine advanced anatomical data with functional data. Examples are the combinations of computerized tomography with positron emission tomography (PET) (for which instrumentation is available), MRI with PET (a prototype instrument is in use, commercial rollout expected in 2010), and diffusion tensor imaging with fMRI. Moreover, the higher resolutions of MRI and computerized tomography are leading to higher-resolution mapping of cortical thickness.

The Army needs to monitor these advanced neuroimaging techniques. Improvements are such that studies that were completed in the past decade or so are being repeated and are yielding much different results than the earlier studies. The use of two or more imaging modalities simultaneously or in sequence offers the exciting prospect of being able to track the dynamics of brain activity on different spatial and temporal scales. To do this, it will be necessary to develop an integrated, dynamic computational framework based on the biophysical, physiological, and anatomical characteristics that can be imaged by these modalities. The modality components of this computational framework could be identified and validated through a series of cross-modal experiments. Some of this work can be done using high-speed computing resources to design and test the dynamic data analysis algorithms on simulated (and, later, experimental) data from multimodal imaging. This cross-modality validation is especially important for the Army in that it directly feeds into the understanding of surrogate measures.

The ability to record information from large numbers of neurons using multielectrode recording techniques, local field potentials, and two-photon imaging now makes it possible to understand in greater detail the functional and anatomical significance of specific brain regions. Similarly, the ability to carry out large-scale simulations of neural models makes it possible to guide experimental research in a principled way using data from experimental measurements to facilitate the choice of model parameters. In this regard, the two arms of computational neuroscience, biophysical and algorithmic, can work in concert with experimental neuroscience at all levels to help integrate information into computational theories of the brain and to validate these theories.

New algorithms will improve quantitative understanding of the information in experimental data. Gaining more insight into how the brain computes will undoubtedly bring new approaches to the design of algorithms for machine learning and computation applicable to a broad range of fields. As an example, a key area for research in neural signal processing will be algorithms to facilitate BMIs. These algorithms must be able to make use of the broad range of neural signals (neural spike trains, local field potentials, electroencephalograph recordings) to control the interactions between humans and machines. This is a very challenging task, because the output of the control strategy—for instance, a particular movement—may be clear, but how the control strategy is represented and carried out in the brain and nervous system is less apparent.

Algorithms must be developed hand in hand with efforts by neurophysiologists to reverse engineer the mechanisms of neural control. It is also important that this algorithm research stays in close contact with the field of control theory, where similar algorithms have been developed to solve problems related to entirely manmade control systems. This research will lead to new algorithms and most likely new theories and practical approaches to the design and implementation of control strategies. Some BMIs might be designed for motor prosthetic purposes, others for control of deep-brain stimulation to treat Parkinson's disease, obsessive compulsive disorder, and depression. Each problem has its unique features and control requirements that will need to be studied in detail to understand how the relevant brain region functions, so that optimal algorithms and, eventually, optimal therapeutic strategies can be devised.

New Types of Averaging in fMRI

Group averages still dominate the literature, with studies utilizing the average activation pattern of 5-10 subjects for a given paradigm. The current trend is to use at least 10-12 subjects and construct a second level of analysis to produce a random-effects average. In a group average, single subjects may dominate data sets and skew the results. In a random-effects average, a single subject's data are treated as a fluctuation from the population average. Another type of analysis is a conjunction of activated areas in a sample of subjects: This type of analysis produces a map based on common (overlapping) regions in each subject's activation map. It is expected that additional methods will emerge that promote understanding of brain function common to all as well as individual and group variations in brain function. This trend should be monitored for possible future Army applications in selection and assessment.

Database Aggregation and Translation for Meta-analyses

Several groups are sponsoring the creation of results databases and proposing standard formats for brain functional and anatomical imaging data, including multimodal techniques. Some are based on cortical surface maps, some on Montreal Neurologic Institute coordinate statistical parametric mapping, and some are based on both. Clearinghouses are under construction for analysis tools (National Institutes of Health Blueprint for Neuroscience Research) and other resources. The Army can leverage these resources for meta-analyses of large data samples to seek out opportunities for further research.

Default Mode Networks

Since the work of Biswal et al. (1997), there has been expanding interest in the so-called default-mode network of the brain. This network is seen to consist of "naturally connected" areas and to include functional connectivity and effective connectivity as well as the difference between the two. The topic is being pursued by those interested in neuroergonomics, and those promoting the topic hypothesize that, ultimately, the efficient use of neural resources takes advantage of these default connections. This work could have implications for cognitive fatigue, learning, and performance optimization. Unlike research in connectomics, this research is noninvasive and is conducted on humans. The Army should monitor this trend for proof that such a default network overlies our physical neural connections.

PRIORITIES FOR ARMY INVESTMENT

The committee was tasked to identify technology development opportunities and to recommend those worthy of investment in the near, medium, and far terms. These technology development opportunities, all of which have been discussed earlier in this chapter, were judged to be "high-priority" (Table 7-1), "priority" (Table 7-2), and "possible future opportunities" (Table 7-3).

The committee asked four questions as it decided which opportunities to include in the tables: Should the Army fund the technology? Should the Army maintain expertise in the technology? Is it likely that the technology, if successful, will have a significant impact? Will there need to be advances in subordinate technologies, such as robust, ruggedized sensors

TABLE 7-1 High-Priority Opportunities for Army Investment in Neuroscience Technologies (Recommendation 14)

Technology Opportunity	ME	RE	Time Frame[a]	Current Investment (L, M, or H) Commercial	Academic
Field-deployable biomarkers of neural state	x	x	Ongoing	L	M
In-helmet EEG for brain–machine interface	x	x	Medium term	M	L
Signal processing and multimodal data fusion, including imaging modalities such as MRI, fMRI, DTI, DSI, PET, and MEG and physiological measures such as heartbeat, interbeat intervals, GSR, optical computer recognition, eye tracking, and pupilometry	x	x	Ongoing	M	H
Soldier models and biomarkers for sleep		x	Ongoing	M	M
Vertical fMRI		x	Medium term	L	L
Fatigue prediction models	x		Medium term	L	M
Behavioral measures of fatigue	x		Medium term	M	L
Prospective biomarkers for predictive measures of soldier response to environmental stress, including hypoxic and thermal challenges	x	x	Medium term	L	L
NIRS/DOT	x	x	Medium term	L	L
Biomedical standards and models for head impact protection, including torso protection from blast	x	x	Medium term	M	M
Threat assessment augmentation	x		Medium term	M	M
fMRI paradigms of military interest		x	Ongoing	L	M

NOTE: ME, mission-enabling; RE, research-enabling; L/M/H, low, medium, or high; EEG, electroencephalography; MRI, magnetic resonance imaging; fMRI, functional magnetic resonance imaging; DTI, diffuse tensor imaging; DSI, diffusion spectrum imaging; PET, positron emission tomography; MEG, magnetoencephalography; NIRS, near-infrared spectroscopy; DOT, diffuse optical tomography; GSR, galvanic skin response.

[a]In this column, "medium term" means between 5 and 10 years and "ongoing" means that results will be available within 5 years, but continuing investment is recommended to stay at the forefront of the technology.

SOURCE: Committee-generated.

TABLE 7-2 Priority Opportunities for Army Investment in Neuroscience Technologies (Recommendation 15)

Technology Opportunity	ME	RE	Time Frame[a]	Current Investment (L, M, or H) Commercial	Academic
Haptic feedback with VR	x		Medium term	H	L
Augmented reality (virtual overlay onto real world)	x	x	Medium term	H	H
In-helmet EEG for cognitive state detection and threat assessment	x	x	Medium term	L	M
Information workload management	x		Far term	L	M
Time-locked, in-magnet VR and monitoring for fMRI		x	Medium term	L	M
Immersive, in-magnet virtual reality		x	Near term	L	M
EEG physiology	x	x	Far term	L	H
Uses of TMS for attention enhancement		x	Medium term	L	M
In-vehicle TMS deployment	x		Far term	L	L
Heartbeat variability	x	x	Near and medium term	L	H
Galvanic skin response	x	x	Near and medium term	H	L

NOTE: ME, mission-enabling; RE, research-enabling; L/M/H, low, medium, or high; VR, virtual reality; TMS, transcranial magnetic stimulation.
[a]In this column, "near term" means within 5 years, "medium term" means between 5 and 10 years, and "far term" means 10-20 years.
SOURCE: Committee-generated.

TABLE 7-3 Possible Future Opportunities (Neuroscience Areas Worthy of Monitoring for Future Army Investment)

Technology Opportunity	ME	RE	Time Frame[a]	Current Investment (L, M, or H) Commercial	Academic
Brain–computer interface system (direct)	x		Far term	H	H
Imaging cognition		x	Far term	L	H
Neuropharmacological technology		x	Far term	M	M
Advanced fMRI data collection		x	Medium term	M	M
Averaging methodology for fMRI		x	Medium term	L	M
Brain database aggregation		x	Far term	M	M
Default mode networks	x	x	Medium term	L	H
Inverse MRI		x	Medium term	L	M
Low-field MRI	x	x	Far term	L	M
Uses of TMS for brain network inhibition		x	Far term	L	M
Safety of multiple exposures to TMS		x	Medium term	M	M
In-helmet TMS deployment	x		Far term	L	L
Connectomics		x	Far term	L	M
Atomic magnetometers	x	x	Far term	M	M

NOTE: ME, mission-enabling; RE, research-enabling; L/M/H, low, medium, or high; fMRI, functional magnetic resonance imaging; MRI, magnetic resonance imaging; and TMS, transcranial magnetic stimulation.
[a]In this column, "medium term" means between 5 and 10 years and "far term" means 10-20 years.
SOURCE: Committee-generated.

and noise-filtering algorithms, before the technology can be implemented?

The committee considered all of the topics in Tables 7-1 and 7-2 worthy of immediate investment but left up to the Army their relative prioritization within each group. Initial priorities might depend, for instance, on the relative importance to the Army of the applications served; these priorities might then change based on research progress. As defined at the very beginning of this chapter, a technology is categorized as mission-enabling (ME column in the tables) if it is instrumental in assisting the warfighter or commander in an operational mission or in a training or assessment mission. It is research-enabling (RE column) if it is instrumental in filling a critical gap in current research capability. Research-enabling instruments are expected to be brought into service on a smaller scale to study and evaluate warfighter or commander performance, perhaps in the laboratory, perhaps in simulated environments. The research is expected to shed neuroscientific light onto current or future Army training and doctrine and to yield concrete suggestions to improve warfighter performance. Note that a technology may be both mission-enabling and research-enabling.

For each opportunity, the Time Frame column gives the committee's estimate of the time needed for development and an idea of when a particular technology will be fielded—that is, the duration of the investment before a product or instrument can be brought into service. The Current Investment column lists the source, academic sector or commercial, and the level of funding being brought to bear on the particular technology in its envisioned Army application. Commercial investment comprises large investments by industry in for-profit ventures. Academic investment comprises investments by various civilian funding agencies such as the National Institutes of Health or the National Science Foundation in university (or other academic) research.

A high (H) level of current investment reflects sufficient external investment to develop the technology to the point where it can be used for Army applications without additional Army investment, although some funding might be required to adapt the technology to a specific application. A medium level of investment (M) reflects funding that would, by itself, allow Army applications to be developed in two or three times the time shown (much slower development than with Army investment in addition to the external sources). A low investment level (L) means that there is little or no investment in Army applications and there will be no technology advance toward an Army application without Army support.

For the high-priority and priority technology development opportunities (Tables 7-1 and 7-2), the committee envisaged the Army employing a mix of internal research and personnel and externally supported research and personnel, as exemplified for research in other fields by Army University Affiliated Research Centers (UARCs), Collaborative Technology Alliances (CTAs), Multidisciplinary University Research Initiatives (MURIs), and faculty fellowships. In addition, as has already been mentioned several times, the committee envisioned the Army would maintain a constant level of expertise to monitor relevant areas of neuroscience research across the board. This would probably entail support from outside experts who would regularly report on progress throughout neuroscience and who might be members of a permanent body established to stay abreast of developments.

Aside from distinguishing between priority and high-priority opportunities, the committee did not prioritize the technology opportunities within a particular table, and all are recommended to receive some level of investment by the Army. The opportunities in Table 7-1 are recommended for initial long-term (5 years or more) commitments. Those in Table 7-2 are recommended for limited (2-3 year initial) commitments to augment the high-priority investments in Table 7-1 and to enable exploration of additional applications that could have a large impact. Their continued funding for longer periods will be guided by evaluations of their research progress. Table 7-3 lists possible future opportunities for consideration by the Army.

In addition to recommending that the Army pursue the listed opportunities, the committee recommends that the Army enhance its existing in-house resources and research capabilities. This would ensure that the Army has mechanisms for interacting with the academic and commercial communities engaged in relevant areas of research and technology development, to monitor progress and decide when future advances in neuroscience technology developments would merit Army investment.

REFERENCES

Biswal, B.B., J. Van Kylen, and J.S. Hyde. 1997. Simultaneous assessment of flow and BOLD signals in resting-state functional connectivity maps. NMR in Biomedicine 10(4-5): 165-170.

Blankenburg, F., C. Ruff, S. Bestmann, O. Josephs, R. Deichman, O. Bjoertomt, and J. Driver. 2008. Right parietal cortex and top-down visuospatial attention: Combined on-line rTMS and fMRI. Presented at HBM2008, the 14th Annual Meeting of the Organization for Human Brain Mapping, Melbourne, Australia, June 15-19.

Bourzac, K. 2008. TR10: Atomic magnetometers. Available at http://www.technologyreview.com/read_article.aspx?ch=specialsections&sc=emerging08&id=20239&a=. Last accessed July 21, 2008.

Cheng, Y., J. Wang, T. Rao, Xi. He, and T. Xu. 2008. Pharmaceutical applications of dendrimers: Promising nanocarriers for drug delivery. Frontiers in Bioscience 13(4): 1447-1471.

Dinges, D.F., M.M. Mallis, G. Maislin, and J.W. Powell. 1998. Evaluation of Techniques for Ocular Measurement as an Index of Fatigue and the Basis for Alertness Management, Report No. DOT HS 808 762. Springfield, Va.: National Technical Information Service.

Genik, R.J., C.C. Green, F.X. Graydon, and R.E. Armstrong. 2005. Cognitive avionics and watching spaceflight crews think: Generation-after-next research tools in functional neuroimaging. Aviation, Space, and Environmental Medicine 76(Supplement 1): B208-B212.

Goldiez, B.F., A.M. Ahmad, and P.A. Hancock. 2007. Effects of augmented reality display settings on human wayfinding performance. IEEE Transactions on Systems, Man, and Cybernetics, Part C: Applications and Reviews 37(5): 839-845.

Goldstone, R.L. 1998 Perceptual learning. Annual Review of Psychology 49: 585-612.

Greene, K. 2007a. Brain sensor for market research: A startup claims to read people's minds while they view ads. Available at http://www.technologyreview.com/Biztech/19833/?a=f. Last accessed July 23, 2008.

Greene, K. 2007b. Connecting your brain to the game: Using an EEG cap, a startup hopes to change the way people interact with video games. Available at http://www.technologyreview.com/Biztech/18276/?a=f. Last accessed July 23, 2008.

Hancock, P.A., and J.S. Warm. 1989. A dynamic model of stress and sustained attention. Human Factors 31: 519-537.

Harris, W.C., P.A. Hancock, and S.C. Harris. 2005. Information processing changes following extended stress. Military Psychology 17(2): 115-128.

Hirshberg, L.M., S. Chiu, and J.A. Frazier. 2005. Emerging brain-based interventions for children and adolescents: Overview and clinical perspective. Child and Adolescent Psychiatric Clinics of North America 14(1): 1-19.

Huettel, S.A., A.W. Song, and G. McCarthy. 2004. Functional Magnetic Resonance Imaging. Sunderland, Mass.: Sinauer Associates, Inc.

Klein, G.A. 1989. Recognition-primed decision. Advances in Man-Machine Systems Research 5: 47-92.

Kraus, R.H., Jr., P. Volegov, A. Matlachov, and M. Espy. 2007. Toward direct neural current imaging by resonant mechanisms at ultra-low field. NeuroImage 39(1): 310-317.

Lee, M., T.T. Chen, M.L. Iruela-Arispe, B.M. Wu, and J.C.Y. Dunn. 2007. Modulation of protein delivery from modular polymer scaffolds. Biomaterials 28(10): 1862-1870.

Lin, F.-H., T. Witzel, J.B. Mandeville, J.R. Polimeni, T.A. Zeffiro, D.N. Greve, G. Wiggins, L.L. Wald, and J.W. Belliveau. 2008. Event-related single-shot volumetric functional magnetic resonance inverse imaging of visual processing. NeuroImage 42(1): 230-247.

Lowe, C. 2008. Land Warrior needs work, soldiers say. Available at http://www.military.com/NewsContent/0,13319,161855,00.html. Last accessed July 23, 2008.

McDermott, R., S.K. Lee, B. ten Haken, A.H. Trabesinger, A. Pines, and J. Clarke. 2004. Microtesla MRI with a superconducting quantum interference device. Proceedings of the National Academy of Sciences of the United States of America 101(21): 7857-7861.

Nature. 2007. Mind games: How not to mix politics and science. Nature 450(7169): 457.

NRC (National Research Council). 1997. Tactical Display for Soldiers: Human Factors Considerations. Washington, D.C.: National Academy Press.

Owen, A.M., R. Epstein, and I.S. Johnsrude. 2002. fMRI: Applications to cognitive neuroscience. Pp. 312-327 in Functional MRI. P. Jezzard, P.M. Matthews, and S.M. Smith, eds. New York, N.Y.: Oxford University Press.

Rogers, T.B., N.A. Kuiper, and W.S. Kirker. 1977. Self-reference and the encoding of personal information. Journal of Personality and Social Psychology 35(9): 677-688.

Scales, R.H. 2006. Clausewitz and World War IV. Available online at http://www.afji.com/2006/07/1866019/. Last accessed July 23, 2008.

Schummers, J., H. Yu, and M. Sur. 2008. Tuned responses of astrocytes and their influence on hemodynamic signals in the visual cortex. Science 320(5883): 1638-1643.

Shepard, R.N., and L.A. Cooper. 1982. Mental Images and Their Transformations. Cambridge, Mass.: MIT Press.

Shepard, R.N., and J. Metzler. 1971. Mental rotation of three dimensional objects. Science 171(3972): 701-703.

Tarr, M.J., and W.H. Warren. 2002. Virtual reality in behavioral neuroscience and beyond. Nature Neuroscience 5(11): 1089-1092.

8

Long-Term Trends in Research

Chapters 3 through 6 identified and discussed near- to medium-term research opportunities that the committee judges could have high value for Army applications. The committee's specific recommendations on these opportunities are presented in Chapter 9. Task 4 of the statement of task asks the committee to "determine trends in research and commercial development of neuroscience technologies that are likely to be of importance to the Army in the longer term." Important long-term trends in technology development were discussed in Chapter 7, along with specific technology opportunities evaluated by the committee. This chapter presents important long-term trends in neuroscience research, including research aimed at expanding our fundamental understanding and applied research that has particular relevance for the Army. The committee believes the Army should monitor the progress of research in these areas and evaluate the results for promising Army-relevant applications. In addition to the research trends themselves, the chapter describes the type of mechanism needed for monitoring progress in both research and technology development.

The soldier is the centerpiece of every Army operation, and the Army will always depend on its soldiers to accomplish its missions. Just as physics and chemistry are the foundational sciences for military ballistics and platform systems, neuroscience as defined in this report is the foundational science for the soldier. The foremost research objectives of the Army should include increasing the survivability of soldiers, both in combat and under the other extreme conditions in which they operate, while sustaining and enhancing their performance.

With these soldier-focused objectives in mind, the committee has identified four trends to represent the breadth and potential importance of research in neuroscience:

- Discovering and validating biomarkers for neural states linked to soldiers' performance outcomes.
- Using individual variability to optimize unit performance.
- Recognizing opportunities from the vertical integration of neuroscience levels.
- Gaining new insights into the behaviors of adversaries.

TREND 1: DISCOVERING AND VALIDATING BIOMARKERS OF NEURAL STATES LINKED TO SOLDIERS' PERFORMANCE OUTCOMES

As discussed in Chapters 3 through 6, the cognitive and behavioral performance of soldiers in many areas—training and learning, decision making, and responding to a variety of environmental stressors—has substantial neurological components. How the brain functions, even how it is functioning at a particular time, makes a difference in these and other types of performance essential to the Army's missions. The techniques used to study and understand brain functioning at all levels—from the molecular and cellular biology of the brain to observable behavior and soldier interactions with other systems—are providing an ever-increasing number of potential indicators of neural status relevant to Army tasks. The Army will need to monitor these techniques and technologies for their potential to serve as biomarkers of differences in neural state that reliably correlate with changes in performance status. To illustrate this tendency for performance biomarkers to emerge from the methods of studying the brain, three broad kinds of such methods are discussed here: genomic and proteomic markers, neuroimaging techniques, and physiological indicators of neural state or behavioral outcome.

Genetic, Proteomic, and Small-Molecule Markers

The development and functioning of the central and peripheral nervous systems of all animals, including humans, are regulated by genomic and proteomic factors. The genomic factors are associated with the nucleic acids of every cell. From embryonic development through senescence, the inherited genome and epigenetic[1] modifications of it regulate the expression of proteins critical for neural cell functions. This regulated gene expression produces signaling elements (transmitters), signal receivers (receptors), guidance of communication processes (axons and dendrites), and cell–cell recognition materials.

Known genetic markers may, for example, allow identification of individuals at greater risk of damage from exposure to chemical agents or more likely to succumb to post-traumatic stress disorder. The cost of genetic tests is likely to decrease substantially in the next decade, while their effectiveness will increase markedly. Of the 20,000-25,000 genes in the human genome, more than 100 are involved in axonal guidance alone (Sepp et al., 2008). At least 89 genes have been shown to be involved in the faulty formation of the axon's myelin sheath (dysmyelination), associated with the development of schizophrenia (Hakak et al., 2001). Understanding the human genes associated with development of the brain and peripheral nervous system can shed light on differential human susceptibilities to brain injury and may aid in predicting which pharmacological agents will be useful for sustaining performance. The Army should position itself to take advantage of the continuing scientific progress in this area.

A proteomic marker (a type of biomarker) is a protein (generally an enzyme) whose concentration, either systemically or in specific tissues, can serve as a reliable and readily measurable indicator of a condition or state that is difficult or even impossible to assay directly. Small variations in gene structure (polymorphisms) are often associated with differences in concentration of a particular protein in a particular tissue of a particular individual, so there are important linkages between genetic factors and proteomic markers. However, specific enzyme concentrations (including tissue-specific concentration) can also be influenced (upregulated or downregulated) in response to environmental factors that vary on timescales of hours, or roughly the timescale of preparation for and conduct of an Army operation. Thus, proteomic markers can vary with recent or current conditions (environmental stressors, for example) and can also reflect the genetic traits of an individual soldier.

Proteomic markers known to signal a change in vigilance or cognitive behavior include salivary amylase, blood homovanillic acid (which correlates with dopamine metabolism), and lactic acid (a metabolic product of glucose metabolism that increases as a result of intense muscle exercise). Proteomic factors associated with fatigue resistance include microtubule-associated protein 2 and the muscarinic acetylcholine receptor. Comparison of an individual's current concentration (titer) of one of these proteomic markers with his or her baseline titer could quantify one or more neural (cognitive/behavioral) states relevant to the status of the individual's current abilities.

Neurohormones and neuropeptides—biologically active molecules much smaller than proteins or the nucleic acids of the genome—are another emerging class of markers of neurological and cognitive state and of psychophysiological response to stress. A study of candidates for the U.S. Navy Sea, Air, and Land Forces (SEALs) found that candidates with strong stress-hormone reactions to behavioral challenges like abrupt changes or interruptions are less likely to complete training successfully than those with weak reactions (Taylor et al., 2006, 2007). Another example is the work discussed in Chapter 3 on oxytocin, a neuropeptide signal, which is released when an individual experiences a sense of trust (Kosfeld et al., 2005; Zak et al., 2005). Hormonal markers are easily gathered with simple blood draws. The level in the bloodstream of a neural signaling molecule such as oxytocin has at best a very indirect relationship to its level in the brain; it may be necessary to figure out how to monitor its release in the hypothalamus. The monitoring of neurohormones and neuropeptides is likely to be a powerful means of identifying individuals who are well suited to particular tasks and may lend itself to assessing candidates for Special Operations training in particular.

Neuroimaging Techniques

Neuroimaging technologies available in the 2008-2010 time frame allow visualization of brain regions that are activated during action-guiding cognitive processes such as decision making. These activation patterns enable brain activity to be correlated with behavior. These imaging technologies and techniques include structural magnetic resonance imaging for volumetric analysis of brain regions, functional magnetic resonance imaging (fMRI) for cognitive control networks, diffusion tensor imaging for transcranial fibers, and hyperspectral electroencephalography (EEG).

Applications to Soldier Training

As an example relevant to evaluation of training, fMRI scans before and after training sessions can be compared to examine changes in the brain's response to novel training-related stimuli. Novel visual and auditory inputs activate the brain in specific regions. An analysis of event-related potentials combined with fMRI before and after novel auditory cues revealed that a particular event-related potential (a P300-like potential, which is to say a positive potential occurring approximately 300 msec after a triggering stimulus) is associated with fMRI patterns of activity in the bilateral foci of the middle part of the superior temporal gyrus (Opitz et al., 1999). Only novel sounds evoke a contrasting event-related potential (an N400-like negative potential). Individuals with a strong response of the second type also have fMRI scans showing activation in the right prefrontal cortex. These

[1]An epigenetic modification refers to changes in gene expression from mechanisms other than alteration of the underlying DNA sequence.

observations suggest that an indicator based on combining fMRI and event-related potential could be used to assess training to criterion. At criterion—for example, when 90 percent of the appropriate responses are exhibited in response to a cue—effective training will no longer elicit a "novel-type" brain functional response or event-related potential response (Opitz et al., 1999).

Fear is a critical response to threat that can compromise appropriate action of an individual soldier or an entire Army unit. To incorporate desensitization to fear-invoking situations into soldier training, fMRI scans could be compared before and after training to determine which environments elicit fear-correlated neural activity patterns. A prime example is the response of soldiers in Operation Desert Shield and Operation Desert Storm when sensors for chemical warfare agents indicated that the environment might contain an active agent. These fear-invoking events led to significant disorganization of military units, even when the sensor warnings were false positives.

Tracking Change in the Visual Field

The ability to track dynamic changes in objects present in a soldier's visual field is of great benefit to Army personnel. Examples include the sudden appearance of a potential threat on a Force XII Battle Command Brigade and Below display and the apparent change of terrain indicating recent placement of an improvised explosive device (IED). Jeremy Wolfe of Harvard has demonstrated that the visual system must focus on only a very limited region within the visual field to detect change (Angier, 2008). To accommodate human limitations, fMRI neurotechnology could be used to detect minor changes in the visual field and correlate them with activation events in the hippocampus (Bakker et al., 2008). Related research has shown that shifts in visual attention to objects in a field of view tend to occur either as a series of microsaccades (rapid naturally occurring eye movements) or in response to cueing signals in the field of view. Recent studies suggest that the latter is more important (Horowitz et al., 2007).

Leveraging Opportunities for Neuroimaging Techniques

EEG and EEG image processing will continue to advance, and EEG will be incorporated in multimodal imaging equipment with magnetic resonance imaging and magnetic encephalography. The high-payoff opportunity here is to leverage this work to develop a sensor array that can be used on a free-moving subject. A good initial goal for proof-of-concept would be the collection of stable trace data from a treadmill runner.

For neuroimaging with near-infrared spectroscopy (NIRS), DARPA has been active in R&D on NIRS sensor arrays that can be worn in situ. This is an opportunity to advance a noninvasive cerebral blood monitoring tool. Expected improvements in the next 5 years include advanced designs for multichannel data collection from cortical sources. In the 10- to 20-year time frame, one R&D opportunity is to use NIRS for more accurate imaging of the deeper brain.

Physiological Indicators of Neural-Behavioral State

Physiological indicators include individual characteristics such as age, gender, muscle power, neuroendocrine effects, neuromuscular function, vascular tone, and circadian cycling. While neural information processing is primarily a result of brain functioning and can be revealed by brain imaging, the general wellness and physiological condition of the entire human organism can affect combat capability and response to threat. This is true in large part because the brain depends on nutrient input (e.g., glucose and oxygen) via the circulatory system and on neuroendocrine function involving other organ systems. (The complex interactions between the brain and other organ systems of the body were discussed in Chapters 2 and 5.)

For Army applications, physiological indicators of neural state are important because they are often more readily accessible and measurable in the field than more direct indicators of neural state derived from neuroimaging techniques. As discussed in Chapter 2 in the section on reliable biomarkers for neurophysiological states and behavioral outcomes and in Chapter 7 in the section on field-deployable biomarkers, the idea is to find a monitorable physiological condition that correlates to a neural state with sufficient accuracy and precision to be useful as a reliable sign of that state. Often, the laboratory studies that define the neural state and establish the correlation will begin with neuroimaging techniques (such as fMRI).

TREND 2: USING INDIVIDUAL VARIABILITY TO OPTIMIZE UNIT PERFORMANCE

Early systems neuroscience experiments used functional neuroimaging tools—fMRI, positron emission tomography, EEG, or magnetoencephalography—to learn how the various brain systems process cognitive and affective functions (Van Horn et al., 2004). It has become increasingly clear that individuals do not process tasks in the same way but instead engage different brain systems, with the particular systems engaged depending in part on the underlying default brain state (Greicius et al., 2003; Esposito et al., 2006). Neuroscientists are increasingly appreciating the importance of interindividual variability regarding which neural signals are operative during various tasks. Instead of being discouraged by this variability, investigators have begun to use its enormous potential for optimizing training and performance. Paying attention to this variability helps in understanding how different individuals learn differently when acquiring a skill or how they organize their behavior when faced with extreme conditions.

How the Individual Variability Insight Affects the Army

The significance for the Army of this paradigm-shifting insight from neuroscience is profound. Understanding in a systematic way the variability among individuals in brain processing when performing the same task is an initial step toward understanding the systematic differences from one individual to another in how they respond in similar circumstances. For example, some individuals engage emotional processing areas even in simple but challenging cognitive tasks (McIntosh et al., 1996; Beauregard et al., 2001). It would be important to determine whether activation of these emotional processing areas helps or interferes with the cognitive task at hand. If it helps in performing the task, perhaps others can learn to enhance their performance on that type of task if they are trained by methods that activate the emotional processing areas. If such activation interferes with the cognitive task, one might implement training strategies that attenuate the effects of emotional processing. Most important, however, whether performance of a cognitive task is enhanced or degraded by emotional processing may differ across groups of individuals. In this case, an optimal training program would develop modulation strategies tailored for a particular group.

The significance of individual variability for optimizing task performance extends to areas other than training and learning. Chapter 4, for example, noted that individuals show traitlike (i.e., stable for the individual over time) differences in decision-making styles. Chapter 5 emphasized the importance of individual variability in neurophysiological responses to stressors typical in Army mission environments and in responses to countermeasures to those stressors. Given that what needs to be optimized from the standpoint of Army operations is performance of the unit, not the individual, the insight from neuroscience is that a higher optimum may be achievable and sustainable by learning to work with and exploit individual variability rather than treating soldiers as interchangeable parts to be mass-produced on a training assembly line.

Neural Correlates for Cultural Differences in Behavior

Even as it comes to accept this insight into the substantial neurological basis for interindividual variability, the Army will still need to look for some general patterns in variability from which to draw statistically valid general conclusions—that is, the conclusions might be expressed in terms of a distribution in a given population or culture rather than as single-point characterizations or predictions. As a simple example, there are cultural differences in attitudes to deception that often depend on the social role of the individual. There are also well-established gender differences in neural correlates of behaviors, such as social responsiveness (Proverbio et al., 2008) and reward anticipation (Hoeft et al., 2008), and some differences are linked to proteomics (e.g., levels of brain-derived neurotrophic factors or other factors expressed in brain tissue). Western and Eastern cultures process mathematics differently (Tang et al., 2006), and there is every reason to expect that such cultural differences will also exist in other cognitive processing tasks. Identifying and understanding these differences will be critical in such applications as interacting with noncombatants in peacekeeping or in security and stability operations (SASO) missions, as well as in predicting adversary responses and decision making.

A still-open question is whether such cultural and individual differences exist all the way down to the genetic/cellular level or whether they disappear at some level. To use the deception example, even if cultural conditioning (or specific training) means that some individuals do not show "normal" physiological or neural indicators of deliberate lying, is there some type of monitoring that could detect if a subject being interrogated is responding in a "contrary-to-truth" manner?

TREND 3: RECOGNIZING OPPORTUNITIES FROM THE VERTICAL INTEGRATION OF NEUROSCIENCE LEVELS

As discussed in Chapter 2, neuroscience exists on four hierarchical levels, which are now being vertically integrated, from the levels of molecules and cells to the levels of behavior and systems. Increasingly, discoveries and advances at one level are leading directly to discoveries and advances at levels higher and lower in the hierarchy. Given this trend, the Army needs to remain cognizant of research in multiple fields that could impact neuroscience applications on multiple levels.

One such opportunity will serve as a simple example: At the level of behavioral research, a variety of methodologies can be used to assess soldier response. Among these are measures of the efficiency with which a task is performed, subjective self-assessments, and psychophysiological correlates. Often the outcomes from these separate methodologies more or less converge. For instance, as an individual's performance of a task degrades, the individual also subjectively feels more stressed by the demands of the task. Convergence of the assessment results increases confidence that we understand the responses. However, what happens when methodologies from the same or different levels in the hierarchy produce results that do not converge? Divergence raises the question of which technique is a better indicator of the subject's real condition, that is, of the "ground truth" (Yeh and Wickens, 1988; Hancock, 1996). Often the question can be answered by moving up or down in the hierarchy of integrated levels to determine, for example, which of the divergent neuroimaging results is consistent with a behavior conventionally used to define a neural state of interest. Or, one might test whether seemingly similar behaviors are truly the same by examining activation patterns from one or more neuroimaging techniques.

Another example of an opportunity derived from the vertical integration hierarchy is this: The neuroscience literature contains many cases in which neurological signals indicate that an attentive brain has processed a signal but the subject, when asked, denies any conscious knowledge of such processing. Are such cases of divergence between assessments also exhibited at different levels—perhaps a general pattern of response divergence—or are they a function of the task that is imposed, of differences in the assessment methods, of characteristics of the individual being tested, or of some other combination of factors?

Divergence of results is likely to become more frequent, at least for a time, as neuroscience assessment methods proliferate. When results from multiple neuroimaging techniques diverge, is this evidence that the physical signal to which a neuroimaging technique is responding is inherently different from the signal to which another technique is responding? Or, is it instead an idiographic characteristic of the individual being tested, or perhaps just an artifact of nuances in the imposed task and the task environment? As they have throughout the history of any scientific discipline, such divergences spur and orient further inquiry into the methodologies themselves and whether the phenomena they transduce as an information signal are in fact correlates of the same underlying reality (here, brain function or processing event). The Army can leverage the ongoing research efforts on neuroscience methods, especially with regard to the method(s) it selects for field implementation. In short, as neuroscience applications come to use two or more neuroimaging techniques, the Army can benefit from the natural direction of inquiry, which is to investigate and resolve seemingly divergent results.

TREND 4: GAINING NEW INSIGHTS INTO ADVERSARY RESPONSE

Negative social interactions, such as the emergence and growth of distrust or the initiation of hostile behavior, are a relevant aspect of social neuroscience. The neural correlates of these types of social behavior are not yet well understood. However, fairly simple interventions at the neural level can have substantial, if temporary, effects on higher-level functioning, such as the administration of thiopental sodium (Pentothal), a medical anesthetic for psychotherapy under sedation in order to recover repressed memories together with the emotion that accompanied the repressed experience.

The work with the neuropeptide oxytocin, described in Chapter 3, indicates that this neural signaling agent is associated with the bonding of offspring and parents and of intimate couples and that it is released under conditions that evoke trust between individuals. Similarly, recent neuroscience experiments have revealed that individuals show specific brain-related changes when engaging in altruistic behavior (Moll et al., 2006). Understanding how the human brain changes as individuals turn to confrontation from cooperation or turn an adversarial relationship into a friendly one provides an important scientific opportunity for the Army in terms of both the overarching implications of such understanding and the concrete mechanisms of neurological changes. Understanding the neural correlates of this change may, for example, help an intelligence analyst to predict when responses of potential adversaries (or of noncombatants potentially allied with the Blue or Red force) will reverse direction.

A MECHANISM FOR MONITORING NEUROSCIENCE RESEARCH AND TECHNOLOGY

Neuroscience research and applications are advancing at a lightning-like pace, and the Army needs to continually assess the potential of these advances. The growing knowledge base will have many direct and indirect applications to soldiers, applications that will increase their operational effectiveness. A neuroscience monitoring group, consisting of recognized leaders in neuroscience research in both the academic and business communities, would help those making Army science and technology funding decisions to assess the relevance of progress in nonmilitary neuroscience to Army applications. Research results and emerging technology can be relevant, whether through direct adaptation for Army use or as a starting point for further Army-oriented R&D, funded or otherwise fostered by the Army. To ensure that this monitoring group remains sensitive to and keeps abreast of Army needs, its membership should include Army civilians and soldiers with appropriate backgrounds and interests to participate meaningfully in the group's deliberations.

The committee envisions that such a monitoring group would operate mainly by attending and reporting on the presentations at conferences and other meetings of professional societies for the neuroscience-relevant disciplines. Of course, the journals of these societies are also important sources of information, but they may not contain ideas, research accomplishments, or information on commercial products, all of which can be gathered by participation in and interactions at professional meetings. Once the nascent and emerging hot topics are identified, the published literature becomes a useful tool for documenting them and covering their progress. The monitoring group would be alert to neuroscience advances and opportunities reported through national and international societies and organizations promoting neuroscience and neuroscience-related research and collaborations of value to the Army. Examples of such organizations include these:

- Cognitive Neuroscience Society,
- Federation of European Neuroscience,
- Human Factors and Ergonomics Society,
- International Society for Magnetic Resonance in Medicine,
- Organization for Human Brain Mapping,
- Radiological Society for North America (focuses on commercial imaging),

- Society for Neuroscience, and
- Society of Automotive Engineers.

There are also regular conferences that the Army can monitor to track the rapidly developing field of computational neuroscience. These include the Neural Information Processing (NIPS) conference, the biannual Statistical Analysis of Neural Data (SAND) workshop, the annual Computational and Systems Neuroscience (Cosyne) meeting, the annual Computational Neuroscience meeting, and the annual Dynamical Neuroscience meeting. The meetings are a venue for discussions on recent work in the fields of computational neuroscience modeling and signal processing.

The number of Army representatives attending a conference should be large enough to allow coverage of simultaneous sessions that may contain relevant research: two or three persons for smaller conferences, six or more for large gatherings. Each attendee could generate a report to the Army and one of them could summarize the advances and identify possible Army applications. At least yearly, the entire monitoring group would gather, together with additional military guests, to share new Army needs and to discuss the group's recent findings and expectations for Army-relevant neuroscience.

REFERENCES

Angier, N. 2008. Blind to change, even as it stares us in the face. New York Times, April 1, 2008, p. F2.

Bakker, A., C.B. Kirwan, M. Miller, and C.E.L. Stark. 2008. Pattern separation in the human hippocampal CA3 and dentate gyrus. Science 319(5870): 1640-1642.

Beauregard, M., J. Levesque, and P. Bourgouin. 2001. Neural correlates of conscious self-regulation of emotion. Journal of Neuroscience 21(18): RC165.

Esposito, F., A. Bertolino, T. Scarabino, V. Latorre, G. Blasi, T. Popolizio, G. Tedeschi, S. Cirillo, R. Goebel, and F. Di Salle. 2006. Independent component model of the default-mode brain function: Assessing the impact of active thinking. Brain Research Bulletin 70(4-6): 263-269.

Greicius, M.D., B. Krasnow, A.L. Reiss, and V. Menon. 2003. Functional connectivity in the resting brain: A network analysis of the default mode hypothesis. Proceedings of the National Academy of Sciences of the United States of America 100(1): 253-258.

Hakak, Y., J.R. Walker, C. Li, W.H. Wong, K.L. Davis, J.D. Buxbaum, V. Haroutunian, and A.A. Fienberg. 2001. Genome-wide expression analysis reveals dysregulation of myelination-related genes in chronic schizophrenia. Proceedings of the National Academy of Sciences of the United States of America 98(8): 4746-4751.

Hancock, P.A. 1996. Effect of control order, augmented feedback, input device and practice on tracking performance and perceived workload. Ergonomics 39(9): 1146-1162.

Hoeft, F., C.L. Watson, S.R. Kesler, K.E. Bettinger, and A.L. Reiss. 2008. Gender differences in the mesocorticolimbic system during computer game-play. Journal of Psychiatric Research 42(4): 253-258.

Horowitz, T.S., E.M. Fine, D.E. Fencsik, S. Yurgenson, and J.M. Wolfe. 2007. Fixational eye movements are not an index of covert attention. Psychological Science 18(4): 356-363.

Kosfeld, M., M. Heinrichs, P.J. Zak, U. Fischbacher, and E. Fehr. 2005. Oxytocin increases trust in humans. Nature 435(7042): 673-676.

McIntosh, A.R., C.L. Grady, J.V. Haxby, L.G. Ungerleider, and B. Horwitz. 1996. Changes in limbic and prefrontal functional interactions in a working memory task for faces. Cerebral Cortex 6(4): 571-584.

Moll, J., F. Krueger, R. Zahn, M. Pardini, R. de Oliveira-Souza, and J. Grafman. 2006. Human fronto-mesolimbic networks guide decisions about charitable donation. Proceedings of the National Academy of Sciences of the United States of America 103(42): 15623-15628.

Opitz, B., A. Mecklinger, A.D. Friederici, and D.Y. von Cramon. 1999. The functional neuroanatomy of novelty processing: Integrating ERP and fMRI results. Cerebral Cortex 9(4): 379-391.

Proverbio, A.M., A. Zani, and R. Adorni. 2008. Neural markers of a greater female responsiveness to social stimuli. BMC Neuroscience 9: 56.

Sepp, K.J., P. Hong, S.B. Lizarraga, J.S. Liu, L.A. Mejia, C.A. Walsh, and N. Perrimon. 2008. Identification of neural outgrowth genes using genome-wide RNAi. Available at http://www.plosgenetics.org/article/info%3Adoi%2F10.1371%2Fjournal.pgen.1000111. Last accessed August 19, 2008.

Tang, Y., W. Zhang, K. Chen, S. Feng, Y. Ji, J. Shen, E.M. Reiman, and Y. Liu. 2006. Arithmetic processing in the brain shaped by cultures. Proceedings of the National Academy of Sciences of the United States of America 103(28): 10775-10780.

Taylor, M., A. Miller, L. Mills, E. Potterat, G. Padilla, and R. Hoffman. 2006. Predictors of success in Basic Underwater Demolition/SEAL (BUD/S) training. Part I: What do we know and where do we go from here? Naval Health Research Center Technical Document No. 06-37.

Taylor, M., G. Larson, A. Miller, L. Mills, E. Potterat, J. Reis, G. Padilla, and R. Hoffman. 2007. Predictors of success in Basic Underwater Demolition/SEAL (BUD/S) training. Part II: A mixed quantitative and qualitative study. Naval Health Research Center Technical Document No. 07-10.

Van Horn, J.D., S.T. Grafton, D. Rockmore, and M.S. Gazzaniga. 2004. Sharing neuroimaging studies of human cognition. Nature Neuroscience 7(5): 473-481.

Yeh, Y.Y., and C.D. Wickens. 1988. Dissociation of performance and subjective measures of workload. Human Factors 30(1): 111-120.

Zak, P.J., R. Kurzban, and W.T. Matzner. 2005. Oxytocin is associated with human trustworthiness. Hormones and Behavior 48(5): 522-527.

9

Conclusions and Recommendations

The previous chapters discussed neuroscience in terms of Army needs and applicable research and technology developments. This chapter presents the committee's conclusions and its recommendations on opportunities in neuroscience for future Army applications.

The applications relating to traditional behavioral sciences spoken of in the statement of task are of known value to the Army. Accordingly, the committee developed 13 recommendations on neuroscience research in these traditional applications. However, because a key driver for the study was to identify high-risk, cutting-edge, high-payoff research, Army support should clearly not be limited to traditional applications. Indeed, the 13 recommendations are followed in this chapter by two recommendations relating to the material in Chapter 7—specifically, to Tables 7-1 and 7-2. Those tables identified nearly two dozen technology opportunities, about half of which were classed as "high priority" and the remainder as simply "priority." Lastly, the committee made two overarching recommendations on what it called cross-cutting issues. Together, the 17 recommendations should help the Army to methodically exploit the expanding realm of neuroscience research.

RECOMMENDATIONS ON NEUROSCIENCE RESEARCH FOR BEHAVIORAL SCIENCE APPLICATIONS

The committee's specific recommendations on research opportunities for applications related to traditional behavioral science are presented in this section under headings corresponding to the first four chapters of the report: training and learning (Chapter 3), optimizing decision making (Chapter 4), sustaining soldier performance (Chapter 5), and improving cognitive and behavioral performance (Chapter 6).

Training and Learning

The Army has long relied on the behavioral and social sciences to guide development and implementation of training. Research capabilities in these fields are limited, and in-house capabilities to perform neuroscience research are minimal.

Conclusion 1. Neuroscience can extend and improve the Army's traditional behavioral science approaches to both training and learning. For example, neuroscience offers new ways to assess how well current training paradigms and accepted assumptions about learning achieve their objectives. Neuropsychological indicators can help assess how well an individual trainee has assimilated mission-critical knowledge and skills. These asssesment tools also will allow the Army to assess individual variability and tailor training regimens to the individual trainee.
Recommendation 1. The Army should adjust its research capabilities to take advantage of the current and emerging advances in neuroscience to augment, evaluate, and extend its approaches to training and learning. Indicators of knowledge and skill acquisition based in neuroscience should be incorporated into the methods of testing for training success. In particular, these indicators should be employed in identifying individual variability in learning and tailoring training regimens to optimize individual learning.

The Army currently relies heavily on broad, general indicators of aptitude to predict training effectiveness and individual success rates. The importance of predicting success rates of soldiers before assigning them to given tasks increases with the cost of training for the task and with the consequences of not performing the task well. In comparison with the indicators that have been developed for assessing how well skills or knowledge have been acquired (Conclusion 1), neurological predictors of soldier performance need much research and development before they will be ready for Army applications.

Conclusion 2. Current methods for characterizing individual capabilities and matching them to the requirements for performing high-cost, high-value Army assignments do not have neuropsychological, psychophysiological, neurochemical, or neurogenetic components. As a first step toward using insights from these and other neuroscience-related fields, results from relatively simple neuropsychological testing could be empirically tested to seek correlations with successful performance in one or more of these high-cost, high value assignments. Those assignment-specific correlations could then be tested for predictive value with subsequent candidates in the same assignment.

Recommendation 2. The Army should investigate neuropsychological testing of candidates for a training course that is already established as a requirement to enter a high-value field. In this way the Army can determine whether an assignment-specific neuropsychological profile can be developed that has sufficiently high predictive value to use in conjunction with established criteria for the assignment. If results for this investigation are positive, the Army should investigate development of assignment-specific profiles for additional assignments.

Optimizing Decision Making

In the past few decades, the boundaries of the behavioral sciences have expanded to incorporate advances in our knowledge of psychology, neuroscience, and economics. These advances cross the hierarchical levels of neuroscience, offering powerful new tools for understanding and improving decision making.

Conclusion 3a. Human decision making is predictably inefficient and often suboptimal, especially when the decisions require assessments of risk and are made under pressure.

Conclusion 3b. Individuals differ in their approach to making decisions. For example, some individuals are more impulsive, while others are more deliberate and less tolerant of risk. These differences do not mean that risk-tolerant individuals are necessarily better or worse decision makers than risk-averse individuals. From an institutional (Army) point of view, different decision-making styles can suit different individuals for different tasks, and different tasks may even require or be better performed by individuals with different decision-making styles. Neuroscience tools are capable of discerning these differences in decision-making style. With enough research, these tools may become capable of discerning neural correlates for the differences.

Recommendation 3. The Army should expand its existing research in behavioral and social sciences to include neuroscience aimed at developing training and assessment tools for decision makers at all levels in the Army.

Sustaining Soldier Performance

The Army generally views the time frame of sustainment in terms of the duration of a single extended operation or action—typically up to 96 hours. In Chapter 5, the committee reviewed neuroscience applications related to understanding, monitoring, and preventing or treating deficits in soldier performance. These deficits may occur during a single extended operation, or, when they are associated with Army concepts such as individual soldier resiliency and unit-level recovery and reset, they can affect performance over longer time frames: weeks, months, and even years. The committee considered prevention interventions not only in the case of events occurring over a day or several days that may be risk factors for acute deficits noticeable immediately, but also in the case of events responsible for longer-term deficits, such as post-traumatic stress disorder (PTSD) and other chronic central nervous system effects of brain trauma.

Individual Variability of Soldiers

In conventional Army operations, a central tenet is to emphasize a common level of operational readiness and performance across individuals as the basis for unit effectiveness rather than individual readiness and performance. Nevertheless, individual soldiers do vary not only in their baseline optimum performance—that is, performance not degraded by sustained stressors—but also in their response to stressors that, on average, cause less-than-optimal performance (performance deficits). In the case of high-value assignments that are very dependent on exceptionally high-performing individuals, such as assignments to Special Operations forces, the Army already acknowledges and takes advantage of individual variability to achieve its objectives.

Conclusion 4. An important lesson from neuroscience is that the ability to sustain and improve performance can be increased by identifying differences in individual soldiers and using individual variability to gauge optimum performance baselines, responses to performance-degrading stressors, and responses to countermeasures to such stressors.

Recommendation 4. To increase unit performance across the full spectrum of operations, the Army should expand its capacity to identify and make use of the individual variability of its soldiers. The Army should undertake R&D and review its training and doctrine to take best advantage of variations in the neural bases of behavior that contribute to performance. In particular, it should seek to understand—and use more widely—individual variability in (1) baseline optimal performance, (2) responses to stressors likely to degrade optimal performance, and (3) responses to countermeasures intended to overcome performance deficits or to interventions intended to enhance performance above an individual's baseline.

Countermeasures to Environmental Stressors

The degradation of a soldier's performance under sustained physical or mental stress is due to both peripheral (e.g., muscular and cardiovascular) systems and the central nervous system (CNS), which are inextricably linked. However, we lack sufficient fundamental understanding of how these systems interact and how they are influenced by the environmental stressors to which a soldier is exposed. For example, physical and mental fatigue are commonly believed to lead to less-than-optimal performance, but neither is well enough characterized or understood to provide a scientific basis for developing countermeasures to both the peripheral and CNS components of the fatigue.

Conclusion 5. Current nutritional countermeasures to fatigue are based primarily on maintaining cardiovascular and muscle function—that is, they counteract physical fatigue—but they fall short of maintaining brain function—that is, they do not counteract mental fatigue. One reason for this is our insufficient understanding of the CNS components of stress-induced degradation of performance.
Recommendation 5. The Army should increase both the pace of and its emphasis on research designed to understand the neural bases of performance degradation under stress, including but not limited to deficits commonly attributed to fatigue, and the interaction of peripheral and CNS factors in responses to stressors. It should apply the results of this research to develop and improve countermeasures such as nutritional supplements and management of sleep/wake and rest/wakefulness cycles.

Sleep is an active process that plays a fundamental role in cognitive functions such as consolidating memory and promoting synaptic plasticity. Prolonged sleep deprivation interferes with these functions and can thus adversely affect performance.

Conclusion 6. Neuroscience is making progress in understanding the essential stages of sleep and their function in normal neural processes. The molecular mechanisms of sleep homeostasis and circadian rhythms are being elucidated. Gene association studies indicate heritable differences in sleep patterns that could be taken into account in predicting how an individual will respond to sleep deprivation.
Recommendation 6. Since many abilities affected by sleep deprivation—vigilance, memory, and perceptual discrimination, for example—are increasingly important elements of soldier performance, the Army should increase its efforts to collaborate with the lead laboratories involved in physiological and molecular research on sleep.

Pharmaceutical Countermeasures to Performance Degradation

Conclusion 7. Advances in neuroscience are enabling the pharmaceutical industry to develop drugs that act on novel targets to affect mood, motivation, memory, and executive function.
Recommendation 7. The Army should establish relationships with the pharmaceutical industry, the National Institutes of Health, and academic laboratories to keep abreast of advances in neuropharmacology, cellular and molecular neurobiology, and neural development and to identify new drugs that have the potential to sustain or enhance performance in military-unique circumstances. However, caution must be exercised to ensure that the benefits outweigh any unforeseen or delayed side effects.

Conclusion 8. Among the neuropharmaceuticals approved by the Food and Drug Administration for specific medical indications, a number have potential off-label uses in sustaining or optimizing performance. However, any compound, natural or synthetic, that acts on the CNS must be assumed, until proven otherwise, to affect multiple neural systems. It is therefore essential that specificity of action be demonstrated. Moreover, the risks of unforeseen or delayed side effects must be considered, particularly before a neuropharmaceutical is widely administered for sustaining or enhancing performance in mission-critical tasks without a specific medical indication to justify its use.
Recommendation 8. Before the Army attempts to employ neuropharmaceuticals for general sustainment or enhancement of soldier performance, the Army should undertake medically informed evidence-based risk-benefit analyses, including performance and clinical measures to assess overall effects, to ensure that the expected benefits of such medication outweigh the risks of negative side effects or delayed effects.

Conclusion 9. The use of new pharmacological agents to restore function, mitigate pain, or otherwise respond to trauma or facilitate recovery from injury or trauma will be a key contribution of neuroscience in the near to medium term. New, highly specific brain receptors have been identified for a number of agents that could have profound effects on the brain and nervous system. The effectiveness of these agents, and the reduction of unwanted systemic side effects, will be enhanced by technologies that target delivery of the pharmacological agent to a specific site. Targeting of drugs to enhance an ability such as situational awareness is technically feasible, but it may be proscribed by societal and ethical norms and is subject to the caveats on pharmacological enhancement of behavior or performance that the committee discussed in Chapters 5 and 6.
Recommendation 9. The Army should support research on novel mechanisms for noninvasive, targeted delivery of pharmacological agents to the brain and nervous system in the course of medical interventions to mitigate the adverse effects of physical injury to the brain or another portion of the nervous system. In the near to medium term, this research should focus on restoring a performance deficit to baseline function rather than enhancing performance beyond that baseline.

Trauma-Induced Stress Disorders, Including Response to Brain Injury

Conclusion 10. Neuroscience research has identified risk factors associated with the development of PTSD and related stress disorders. The evidence is increasing that these stress disorders are more common among soldiers than was formerly believed.
Recommendation 10. The Army should support continued research on the identification of risk factors for the development of post-traumatic stress disorder (PTSD). This research could inform interventions that mitigate the risk for PTSD and related stress disorders, thereby lessening the performance deficits and disability resulting from these disorders.

Conclusion 11. A blast may cause unique brain injuries resulting in persistent deleterious effects on mood, motivation, and cognition. These effects are likely to undermine the resilience of the soldiers who experience them by degrading their performance and to detract from the unit morale that is so essential for effective reset and recovery following combat operations. Although improved protective materials and headgear (helmet) configurations might help, body armor improvements seem unlikely to resolve most of the persistent neurological effects. Medication and other neurophysiological remediation, starting with immediate postblast care, are likely to have a much more profound effect than further improvements to body and head armor.
Recommendation 11. The Army should apply the rapidly advancing understanding of the acute neuropathology of blast-induced traumatic brain injury, including the delayed neuropsychiatric effects of injuries as well. Mitigation strategies should include immediate postblast care using medication and/or other neuroprotective approaches proven to reduce the risk and severity of performance degradation. The Army should also continue its research in protective body armor.

Improving Cognitive and Behavioral Performance

Conclusion 12. Increased vigilance and enhanced perceptual discrimination, such as being able to recognize salient features or patterns, are inherently valuable to military missions. Research in a number of neuroscience subdisciplines, including computational neuroscience, systems neuroscience, and neuroergonomics, could lead to significant improvements in the skills and capabilities of soldiers and officers along these lines.
Recommendation 12. The Army should structure its announcements of opportunities for research to draw broadly on multiple scientifically sound approaches to improving cognitive and behavioral performance, extending across the entire spectrum of neuroscience research rather than relying on a single approach. Army research opportunities should foster peer-reviewed competition and the synergism of collaboration across subdisciplines and approaches.

Conclusion 13. Neuroergonomics, an emerging field within the broader field of brain–machine interfaces, explores the ability of the brain to directly control systems by means that go beyond the usual human effector system (hands and the voice). This is accomplished by structuring the brain's output as a signal that can be transduced into an input to an external machine, electronic system, computer, semiautonomous air or ground vehicle, and the like. The Army Research Laboratory is now exploring the potential benefits of neuroergonomics. In the Army context, the goal of neuroergonomics is to facilitate a soldier–system symbiosis that measurably outperforms conventional human–system interfaces.
Recommendation 13. The Army should continue its focus on neuroergonomic research, using measured improvements in performance over selected conventional soldier–system interfaces as the metric to evaluate the potential of neurophysiology and other neuroscience disciplines in Army-relevant R&D for improving cognitive and behavioral performance.

RECOMMENDATIONS ON NEUROSCIENCE TECHNOLOGY DEVELOPMENT

In Chapter 7, the committee identified and assessed cutting-edge, high-payoff technology opportunities, emphasizing their potential value for Army applications. Technologies were evaluated first with respect to their ability to enable Army missions (these were called mission-enabling technologies) and then with respect to their ability to support neuroscience research of high relevance to Army applications (these were called research-enabling). Sometimes, a technology is both mission enabling and research enabling, but in all cases, it must be capable of being scientifically validated.

To arrive at opportunities it could recommend for Army investment, the committee considered not only the potential value of the technologies to the Army but also the time frame for developing an initial operational capability and the extent of external investment that the Army could leverage. In this way it came up with a set of "high-priority" technology opportunities it believed would have the greatest potential for high payoffs in Army applications and best deserved Army investment. It also identified a second set of "priority" opportunities that could augment the first set, and a third set whose progress should be monitored for future consideration.

Conclusion 14. Table 7-1 lists the set of opportunities in neuroscience technology development that the committee believes are a high priority for Army investments. It is critical that the emerging technology development pursued by the Army be subjected to rigorous scientific and operational validation.

Recommendation 14. The Army should invest in the high-priority technology opportunities listed in Table 7-1. The investments should initially include long-term (5 or more years) commitments to each opportunity.

Conclusion 15. Table 7-2 lists additional opportunities in neuroscience that the committee recommends for Army investment. The committee views these opportunities as supplementing those in Table 7-1 and as deserving of somewhat less R&D funding, to at least explore their potential applications.
Recommendation 15. The Army should consider limited investments (2 or 3 years for the initial commitment) in the technology opportunities listed in Table 7-2. Evaluation of the results for each initial investment combined with assessment of outside progress in the field should guide decisions on whether to continue the funding for additional periods.

OVERARCHING RECOMMENDATIONS

The preceding 15 recommendations respond directly to one or more items in the statement of task for the committee. In reflecting on the feasibility of actually implementing these recommendations, the committee found two crosscutting issues that go beyond any particular request in the statement of task but that the Army must address if the potential value of neuroscience is to be tapped in a substantial way.

A Mechanism for Monitoring New Opportunities in Neuroscience Research and Technology

Neuroscience is growing rapidly as discoveries in multiple fields are linked to our expanding knowledge and understanding of brain functions. This expansion of neuroscience applications in multiple areas of importance to the Army has led to a division of responsibilities for developing objectives and implementing neuroscience research among multiple organizations. A more serious problem is that there is currently no single point in the Army science and technology structure where progress in neuroscience, construed broadly, is being monitored for potential Army applications and from which coordinating guidance can be disseminated to the distributed centers of relevant neuroscience-based R&D. The committee views this lack of focus on identifying and leveraging the rapid advances in neuroscience, together with the dispersion of largely isolated R&D activities, as the most significant barrier to implementation of the specific recommendations presented above.

In addition to the specific technology development opportunities for Army investment listed in Tables 7-1 and 7-2, the committee identified future opportunities where external progress in neuroscience R&D needs to be monitored by the Army (Table 7-3). The committee also identified four trends in neuroscience research likely to one day yield opportunities of great benefit to the Army:

- Discovering and validating biomarkers for neural states linked to soldiers' performance outcomes.
- Using individual variability to optimize unit performance.
- Recognizing opportunities from the vertical integration of neuroscience levels.
- Gaining new insights into the behaviors of adversaries.

These pursuits for research will continue to revolutionize our understanding of the embodied mind and foster practical applications in civilian, commercial, and military affairs.

Conclusion 16. Neuroscience research and applications are advancing at a lightning pace. To assess on a continuing basis the potential of these advances on many fronts and to make sound decisions for funding priorities based on them, the Army needs a reliable way to monitor progress in areas of nonmilitary neuroscience research and technology development. Direct Army investment in these areas will probably not be warranted unless an Army-unique application of substantial value emerges. Nonetheless, the Army should stay abreast of what is happening and have mechanisms in place to leverage the research results and adapt new technology to Army applications.
Recommendation 16. The Army should establish a group consisting of recognized leaders in neuroscience research in both the academic and private sectors to track progress in nonmilitary neuroscience R&D that could be relevant to Army applications. To ensure that the monitoring group remains sensitive to and abreast of Army needs, the membership should also include Army civilians and soldiers whose backgrounds and interests would suit them for meaningful participation in the group's deliberations.

Individual Variability as a Future Force Multiplier

A number of conclusions drawn by the committee (Conclusions 2, 3b, and 4) and Long-Term Trend 2, discussed in Chapter 8, relate to a common theme emerging from current neuroscience research—namely, that individual differences in behavior, cognition, and performance of skilled tasks are as deeply rooted in the neural structure of individuals as differences in strength, stamina, height, or perceptual acuity are rooted in physiology. This common theme offers great opportunity to the future Army.

Conclusion 17. Neuroscience is establishing the role that neural structures play in the individual variability observed in cognition, memory, learning behaviors, resilience to stressors, and decision-making strategies and styles. Differences from one soldier to the next have consequences for most of the Army applications discussed in this report. Individual variability influences operational readiness and the ability of military units to perform assigned tasks optimally, but it is in many ways at odds with the conventional approach of training soldiers to be interchangeable components of a unit.

Recommendation 17. Using insights from neuroscience on the sources and characteristics of individual variability, the Army should consider how to take advantage of variability rather than ignoring it or attempting to eliminate it from a soldier's behavior patterns in performing assigned tasks. The goal should be to seek ways to use individual variability to improve unit readiness and performance.

Appendixes

Appendix A

Biographical Sketches of Committee Members

Floyd E. Bloom (NAS/IOM), *Chair*, is professor emeritus in the Department of Molecular and Integrative Neuroscience at the Scripps Research Institute. He has served on numerous committees at the National Academies, including the Committee on Publications and the Symposium on Neuroscience and Brain Research. He is former co-chair of the Report Review Committee and is former chair of the selection committee for the Award in the Neurosciences. He received an M.D. from Washington University and a B.A. from Southern Methodist University in Dallas, Texas.

Richard A. Andersen (NAS/IOM) is professor of neuroscience at the California Institute of Technology in Pasadena. He pioneered the study of brain processes for sight, hearing, balance, and touch; the neural mechanisms of action; and the development of neural prosthetics. Dr. Andersen was awarded the McKnight Technical Innovation in Neuroscience Award in 2000 and the McKnight Neuroscience Brain Disorders Award in 2005. He is an expert in the field of brain–computer interfaces and has many publications on his work. Dr. Andersen earned a Ph.D. at the University of California at San Francisco, and he served as a postdoctoral fellow at the Johns Hopkins Medical School in Baltimore. He earned a B.S. in biochemistry from the University of California at Davis.

Ronald R. Blanck is vice chairman of Martin, Blanck & Associates, a health-care consulting firm serving the private sector and government. He retired as the U.S. Army surgeon general in 2000 and as the president of the Health Science Center at the University of North Texas in 2007. Lt. Gen. (ret.) Blanck commanded the U.S. Army Medical Command and was a battalion surgeon in the Vietnam War. His military honors include the Distinguished Service Medal, the Defense Superior Service Medal, the Legion of Merit, and the Bronze Star Medal. He has consulted for the Army Medical Research and Materiel Command and taught at Georgetown University, George Washington University, Howard University School of Medicine, the Uniformed Services University, the University of Texas Health Science Center at San Antonio, and the University of North Texas Science Center. He is a graduate of the Philadelphia College of Osteopathic Medicine and is board certified in internal medicine.

Emery N. Brown (IOM) is professor of anesthesia in the Department of Anesthesia and Critical Care at Massachusetts General Hospital. He is currently also serving as a member of the Committee to Evaluate the National Science Foundation Vertically Integrated Grants for Research and Education Program and recently served on the Committee on Applied and Theoretical Statistics. Dr. Brown's expertise has earned him many awards and honors, including the Harvard Medical School/Hewlett-Packard Outstanding Medical School Graduate Award and the National Institute of Mental Health Independent Scientist Award. He is an elected fellow of the American Institute for Medical and Biological Engineering. He earned an M.D. magna cum laude, a Ph.D. in statistics, and an M.A. in statistics from Harvard University. He earned a B.A. in applied mathematics magna cum laude from Harvard College.

Joseph T. Coyle (IOM) is the Eben S. Draper Professor of Psychiatry and Neuroscience at Harvard University. He was Distinguished Service Professor of Child Psychiatry at the Johns Hopkins University School of Medicine, chair of the Consolidated Department of Psychiatry at Harvard Medical School, and president of the American College of Neuropsychopharmacology and the Society for Neuroscience. Dr. Coyle's many awards and honors include the Gold Medal Award from the Society of Biological Psychiatry and the Pasarow Foundation Award for research in neuropsychiatry. He graduated in cursa honoris cum laude from the College of the Holy Cross and earned an M.D. from the Johns Hopkins University School of Medicine.

Mary L. (Missy) Cummings is an assistant professor in the Aeronautics and Astronautics Department at the Massachusetts Institute of Technology. She performs research in collaborative human–computer decision making for command and control domains and is a recognized expert in the area of human superviory control. Dr. Cummings graduated from the U.S. Naval Academy with a B.S. in mathematics in 1988. She received an M.S. in space systems engineering from the Naval Postgraduate School in 1994 and a Ph.D. in systems engineering from the University of Virginia in 2003. Dr. Cummings served as a naval officer from 1988 to 1999 and was among the first female fighter pilots in the Navy.

J. Mark Davis is professor and director of the Psychoneuroimmunology, Exercise and Nutrition Laboratory in the Division of Applied Physiology within the Department of Exercise Science at the University of South Carolina. He has published over 110 peer-reviewed articles of relevance to soldier nutrition, physical training, and mental/physical performance, including "Possible mechanisms of central nervous system fatigue during exercise" and "Effects of branched-chain amino acids and carbohydrate on fatigue during intermittent, high-intensity running." He earned a Ph.D. from Purdue University and a B.S. from California Polytech in San Luis Obispo.

Michael S. Gazzaniga (IOM) is the first director of the Sage Center for the Study of Mind at the University of California, Santa Barbara. He is past director of the Center for Neuroscience at the University of California, Davis. Through his extensive work with split-brain patients, Dr. Gazzaniga has made important advances in our understanding of functional lateralization in the human brain and of how the cerebral hemispheres communicate with one another. His research is well known in both clinical and basic science circles, and he has written several highly acclaimed books, including the landmark 1995, 2000, and 2004 editions of *The Cognitive Neurosciences*, which is recognized as the sourcebook in the field. Dr. Gazzaniga is the president of the Cognitive Neuroscience Institute, which he founded, and is the editor in chief emeritus of the *Journal of Cognitive Neuroscience*. Dr. Gazzaniga was elected to the American Academy of Arts and Sciences. He is the elected president of the American Psychological Society, and he also serves on the President's Council on Bioethics. He received a Ph.D. in psychobiology from the California Institute of Technology. He will be delivering the Gifford Lectures at the University of Edinburgh in 2009.

Richard J. Genik, II, is director of the Emergent Technology Research Division and research assistant professor in the Department of Psychiatry and Behavioral Neurosciences at Wayne State University. Among his many areas of expertise are the use of magnetic resonance imaging and the use of fMRI to measure cognitive workload in naturalistic, multitasking environments. Dr. Genik has 131 peer-reviewed publications, including "Watching people think" and "Scientific methods that may predict behaviors," which appeared in *Biotechnology Trends Relevant to Warfare Initiatives* in 2005. He has a Ph.D. in physics from Michigan State University and a B.S. in applied physics from Wayne State University.

Paul W. Glimcher is professor of neural sciences, economics, and psychology at New York University's Center for Neural Science and is the director of the Center for Neuroeconomics at New York University. He has achieved the following: A.B., Princeton University, magna cum laude; Ph.D., University of Pennsylvania, neuroscience; fellow of the McKnight, Whitehall, Klingenstein, and McDonnell foundations. Dr. Glimcher is an investigator for the National Eye Institute, the National Institute of Mental Health, and the National Institute of Neurological Disorders and Stroke. He was the founding president of the Society for Neuroeconomics; winner of the Margaret and Herman Sokol Faculty Award in the Sciences, 2003; and winner of NYU's Distinguished (Lifetime Accomplishment) Teaching Award, 2006. He has had articles published in *Nature*, *Science*, *Neuron Journal of Neurophysiology*, *American Economic Review*, *Games and Economic Behavior, Vision Research*, *Experimental Brain Research*, and the *MIT Encyclopedia of Cognitive Science* and was international author of *Decisions, Uncertainty, and the Brain: The Science of Neuroeconomics* from MIT Press and winner of the American Association of Publishers Medical Sciences Book of the Year, 2003. Professor Glimcher's work has been covered by *The Wall Street Journal*, *Time*, *Newsweek*, *The Los Angeles Times*, *Money Magazine*, and *The New Scientist* and featured on National Public Radio, the BBC, and Fox News, among others.

Peter A. Hancock is provost distinguished research professor for the Department of Psychology, the Institute for Simulation and Training, and the Department of Civil and Environmental Engineering at the University of Central Florida in Orlando. He is an expert in human factors and ergonomics and serves on the National Research Council's Committee on Human Factors. Among his many awards, he received the John C. Flanagan Award from the Society of Military Psychologists of the American Psychological Association in 2007. Dr. Hancock has authored over 500 refereed scientific articles and publications, including the handbook on perception and cognition *Human Performance and Ergonomics*. He earned a D.Sci. as well as a E.Ed. in anatomy and physiology at Loughborough University in Loughborough, England. He also earned a Ph.D. at the University of Illinois at Champaign.

Steven Kornguth is director of the Center for Strategic and Innovative Technologies and a professor of pharmacy at the University of Texas at Austin. He is a member of the American Society of Neurochemistry, the Neuroscience Society, and the Army Science Board. Dr. Kornguth is professor emeritus, neurology and biomolecular chemistry, at the University of Wisconsin, Madison. He served on the U.S. Army Research Laboratory Biotechnology Assessment Committee. He co-organized a joint U.S. Army–Israeli Ministry of Defense conference on bioremediation and has published on many neuroscience topics, including the structure of human synaptic complexes. He earned a B.A. from Columbia University in New York City and a Ph.D. in biochemistry from the University of Wisconsin, Madison.

Martin P. Paulus is professor in residence for the Department of Psychiatry at the University of California at San Diego. Dr. Paulus has a number of publications, including "A temporal and spatial scaling hypothesis for the behavioral effects of psychostimulants," published in 1991, and "A realistic, minimal 'middle layer' for neural networks," published in 1989. He received the Society for Biological Psychiatry Outstanding Resident Award in 1997 and the National Alliance for the Mentally Ill Young Investigator Award in 2000. He earned his doctorate in medicine at Johannes Gutenberg University, Mainz, Germany. He is a member of the American College of Neuropsychopharmacology and a member of the Society for Neuroscience.

Judith L. Swain (IOM) is executive director of the Singapore Institute for Clinical Sciences, Lien Ying Chow Professor of Medicine at the National University of Singapore Yong Loo Lin School of Medicine, and adjunct professor at the University of California at San Diego. A proven leader, Dr. Swain served as chair of the Department of Medicine at Stanford University, President of the American Society for Clinical Investigation and is currently president of the American Association of Physicians. She is a member of the Institute of Medicine (IOM) and currently serves on the IOM Council and the National Research Council's Board on Army Science and Technology. She earned an M.D. from the University of California at San Diego and a B.S. in chemistry from the University of California at Los Angeles.

Paul J. Zak is professor of economics and founding director of the Center for Neuroeconomics Studies at Claremont Graduate University. Dr. Zak also serves as professor of neurology at Loma Linda University Medical Center and is a senior researcher at the University of California at Los Angeles. He is credited with the first published use of the term "neuroeconomics" and has been at the vanguard of this new discipline, which integrates neuroscience and economics. He organized the world's first doctoral program in neuroeconomics at Claremont Graduate University and now administers it. Dr. Zak's lab discovered in 2004 that a chemical in our brains, oxytocin, allows us to determine whom to trust. This knowledge is being used to understand the basis for modern civilizations and economies, for negotiating, and for treating patients with neurological and psychiatric disorders. He has degrees in mathematics and economics from San Diego State University, a Ph.D. in economics from the University of Pennsylvania, and postdoctoral training in neuroimaging from Harvard.

Appendix B

Committee Meetings

FIRST MEETING, DECEMBER 17-18, 2007, WASHINGTON, D.C.

Neuroscience for Future Army Applications—Human Dimension Implications
BG Peter Palmer, U.S. Army Capabilities Integration Center

Army Science and Technology
Thomas Killion, Deputy Assistant Secretary of the Army (Research and Technology)

ARL Research in Neuroscience
Kaleb McDowell, Army Research Laboratory

Medical Research and Materiel Command Research in Neuroscience
Thomas Balkin, Walter Reed Army Institute of Research

SECOND MEETING, FEBRUARY 12-13, 2008, IRVINE, CALIFORNIA

Decision Under Uncertainty
Craig Fox, University of California at Los Angeles

Affective Decision Making
Antoine Bechara, University of Southern California, College of Letters, Arts, and Sciences

Opportunities Offered by Bayesian Decision Theory Applied to Human Movement
Konrad Kording, Northwestern University

Learning in Decision Making: Army Opportunities
Nathaniel Daw, New York University

The Neuroscience of Virtual Humans
Jonathan Gratch, Institute of Creative Technologies

Virtual Reality Applications for Motor Rehabilitation
Albert "Skip" Rizzo, Institute of Creative Technologies

Dynamical Models of Decision Making
James L. McClelland, Stanford University

Electrophysiological Measures and Cognitive Efficiency
Steven A. Hillyard, University of California at San Diego

Neuroscience of Cognition Advances and Opportunities
G. Ron Mangun, University of California at Davis

VRPSYCH Lab: Virtual Reality and Neuropsychology
Thomas D. Parsons, Institute of Creative Technologies

THIRD MEETING, APRIL 28-29, 2008, WASHINGTON, D.C.

ARI Overview Brief
Michelle Sams, U.S. Army Research Institute for the Behavioral and Social Sciences (ARI)

Technology Watch Functional Brain Imaging, ARO MURI 2008, Expertise in Online Forums, Basic Research at ARI
Joseph Psotka, ARI

Selective Instruments for Flight Training
Larry Katz, ARI

Identifying Experts in the Detection of IED
Jennifer Murphy, ARI

Telemedicine and Advanced Technology Research Center, Critical Military Medical Technologies
Col Carl Friedl, Army Medical Research and Materiel Command

FOURTH MEETING, JUNE 29-30, 2008, WASHINGTON, D.C.

Briefing to NRC Neuroscience Meeting
Amy Kruse, Defense Advanced Research Projects Agency

Appendix C

Sampling of Behavioral and Neuropsychological Literature (2001-2007) on High-Performance Athletes

The numbered references in this list are keyed to the discussion in Chapter 3 of topics in the extensive literature on behavioral and cognitive-neural aspects of athletes. Committee members also recommended several older studies or reviews (published prior to 2001) that are important sources for a subject area.

TRAINING METHODS, PERFORMANCE EVALUATION, SOCIAL INTERACTIONS, AND PSYCHOLOGICAL ISSUES

1. Abbott A, Collins D. Eliminating the dichotomy between theory and practice in talent identification and development: considering the role of psychology. J Sports Sci. 2004 May;22(5):395-408.
2. Abrahamsen FE, Roberts GC, Pensgaard AM. An examination of the factorial structure of the Norwegian version of the sport anxiety scale. Scand J Med Sci Sports. 2006 Oct;16(5):358-63.
3. Allami N, Paulignan Y, Brovelli A, Boussaoud D. Visuo-motor learning with combination of different rates of motor imagery and physical practice. Exp Brain Res. 2008 Jan;184(1):105-13.
4. Allegre B, Noel-Jorand MC, Souville M, Pellegrin L, Therme P. Intensive physical activity and alexithymia: results from swimmers' discourse analysis. Psychol Rep. 2007 Jun;100(3 Pt 2):1129-39.
5. Andersen MB, Van Raalte JL, Brewer BW. Sport psychology service delivery: staying ethical while keeping loose. Prof Psychol Res Pr. 2001 Feb;32(1):12-8.
6. Armstrong LE, VanHeest JL. The unknown mechanism of the overtraining syndrome: clues from depression and psychoneuroimmunology. Sports Med. 2002;32(3):185-209.
7. Bach P, Tipper SP. Bend it like Beckham: embodying the motor skills of famous athletes. Q J Exp Psychol (Colchester). 2006 Dec;59(12):2033-9.
8. Backmand H, Kaprio J, Kujala U, Sarna S. Personality and mood of former elite male athletes--a descriptive study. Int J Sports Med. 2001 Apr;22(3):215-21.
9. Backmand H, Kaprio J, Kujala U, Sarna S. Influence of physical activity on depression and anxiety of former elite athletes. Int J Sports Med. 2003 Nov;24(8):609-19.
10. Backmand H, Kaprio J, Kujala UM, Sarna S, Fogelholm M. Physical and psychological functioning of daily living in relation to physical activity. A longitudinal study among former elite male athletes and controls. Aging Clin Exp Res. 2006 Feb;18(1):40-9.
11. Beedie CJ, Coleman DA, Foad AJ. Positive and negative placebo effects resulting from the deceptive administration of an ergogenic aid. Int J Sport Nutr Exerc Metab. 2007 Jun;17(3):259-69.
12. Bekiari A, Patsiaouras A, Kokaridas D, Sakellariou K. Verbal aggressiveness and state anxiety of volleyball players and coaches. Psychol Rep. 2006 Oct;99(2):630-4.
13. Brenner JS. Overuse injuries, overtraining, and burnout in child and adolescent athletes. Pediatrics. 2007 Jun;119(6):1242-5.
14. Bullock N, Martin DT, Ross A, Rosemond D, Marino FE. Effect of long haul travel on maximal sprint performance and diurnal variations in elite skeleton athletes. Br J Sports Med. 2007 Sep;41(9):569-73; discussion 73.
15. Carr CM. Sport psychology: psychologic issues and applications. Phys Med Rehabil Clin N Am. 2006 Aug;17(3):519-35.
16. Chatzisarantis NL, Hagger MS. The moral worth of sport reconsidered: contributions of recreational sport and competitive sport to life aspirations and psychological well-being. J Sports Sci. 2007 Jul;25(9):1047-56.
17. Conroy DE. Representational models associated with fear of failure in adolescents and young adults. J Pers. 2003 Oct;71(5):757-83.
18. Coutts AJ, Wallace LK, Slattery KM. Monitoring changes in performance, physiology, biochemistry, and psychology during overreaching and recovery in triathletes. Int J Sports Med. 2007 Feb;28(2):125-34.
19. Cresswell S, Hodge K. Coping skills: role of trait sport confidence and trait anxiety. Percept Mot Skills. 2004 Apr;98(2):433-8.
20. Cumming J, Olphin T, Law M. Self-reported psychological states and physiological responses to different types of motivational general imagery. J Sport Exerc Psychol. 2007 Oct;29(5):629-44.
21. Daniels AM. Cooperation versus competition: is there really such an issue? New Dir Youth Dev. 2007 Fall(115):43-56, 6-7.
22. Degoutte F, Jouanel P, Begue RJ, Colombier M, Lac G, Pequignot JM, et al. Food restriction, performance, biochemical, psychological, and endocrine changes in judo athletes. Int J Sports Med. 2006 Jan;27(1):9-18.
23. Demorest RA, Landry GL. Training issues in elite young athletes. Curr Sports Med Rep. 2004 Jun;3(3):167-72.
24. Di Russo F, Taddei F, Apnile T, Spinelli D. Neural correlates of fast stimulus discrimination and response selection in top-level fencers. Neurosci Lett. 2006 Nov 13;408(2):113-8.
25. Dietrich A, Sparling PB. Endurance exercise selectively impairs prefrontal-dependent cognition. Brain Cogn. 2004 Aug;55(3):516-24.
26. Dionigi R. Competitive sport and aging: the need for qualitative sociological research. J Aging Phys Act. 2006 Oct;14(4):365-79.

27. Donohue B, Covassin T, Lancer K, Dickens Y, Miller A, Hash A, et al. Examination of psychiatric symptoms in student athletes. J Gen Psychol. 2004 Jan;131(1):29-35.
28. Donohue B, Silver NC, Dickens Y, Covassin T, Lancer K. Development and initial psychometric evaluation of the sport interference checklist. Behav Modif. 2007 Nov;31(6):937-57.
29. Erlanger D, Feldman D, Kutner K, Kaushik T, Kroger H, Festa J, et al. Development and validation of a web-based neuropsychological test protocol for sports-related return-to-play decision-making. Arch Clin Neuropsychol. 2003 Apr;18(3):293-316.
30. Everhart B, Kernodle M, Ballard K, McKey C, Eason B, Weeks M. Physical activity patterns of college students with and without high school physical education. Percept Mot Skills. 2005 Jun;100(3 Pt 2):1114-20.
31. Filaire E, Sagnol M, Ferrand C, Maso F, Lac G. Psychophysiological stress in judo athletes during competitions. J Sports Med Phys Fitness. 2001 Jun;41(2):263-8.
32. Finaud J, Degoutte F, Scislowski V, Rouveix M, Durand D, Filaire E. Competition and food restriction effects on oxidative stress in judo. Int J Sports Med. 2006 Oct;27(10):834-41.
33. Fontani G, Lodi L, Felici A, Migliorini S, Corradeschi F. Attention in athletes of high and low experience engaged in different open skill sports. Percept Mot Skills. 2006 Jun;102(3):791-805.
34. Fung L. Task familiarity and task efficacy: a study of sports coaches. Percept Mot Skills. 2002 Oct;95(2):367-72.
35. Garcin M, Billat V. Perceived exertion scales attest to both intensity and exercise duration. Percept Mot Skills. 2001 Dec;93(3):661-71.
36. Garcin M, Mille-Hamard L, Billat V. Influence of aerobic fitness level on measured and estimated perceived exertion during exhausting runs. Int J Sports Med. 2004 May;25(4):270-7.
37. Garcin M, Mille-Hamard L, Duhamel A, Boudin F, Reveillere C, Billat V, et al. Factors associated with perceived exertion and estimated time limit at lactate threshold. Percept Mot Skills. 2006 Aug;103(1):51-66.
38. Gentner NB, Fisher LA, Wrisberg CA. Athletes' and coaches' perceptions of sport psychology services offered by graduate students at one NCAA Division I university. Psychol Rep. 2004 Feb;94(1):213-6.
39. Glick ID, Horsfall JL. Diagnosis and psychiatric treatment of athletes. Clin Sports Med. 2005 Oct;24(4):771-81, vii.
40. Hale BS, Koch KR, Raglin JS. State anxiety responses to 60 minutes of cross training. Br J Sports Med. 2002 Apr;36(2):105-7.
41. Hamer M, Karageorghis CI, Vlachopoulos SP. Motives for exercise participation as predictors of exercise dependence among endurance athletes. J Sports Med Phys Fitness. 2002 Jun;42(2):233-8.
42. Hannon B, Fitzgerald P. Anxiety and performance in elite non-professional athletes. Ir Med J. 2006 Sep;99(8):238-40.
43. Hanton S, Fletcher D, Coughlan G. Stress in elite sport performers: a comparative study of competitive and organizational stressors. J Sports Sci. 2005 Oct;23(10):1129-41.
44. Hatzigeorgiadis A, Biddle SJ. Cognitive interference during competition among volleyball players with different goal orientation profiles. J Sports Sci. 2002 Sep;20(9):707-15.
45. Hodges NJ, Kerr T, Starkes JL, Weir PL, Nananidou A. Predicting performance times from deliberate practice hours for triathletes and swimmers: what, when, and where is practice important? J Exp Psychol Appl. 2004 Dec;10(4):219-37.
46. Hunt TN, Ferrara MS, Miller LS, Macciocchi S. The effect of effort on baseline neuropsychological test scores in high school football athletes. Arch Clin Neuropsychol. 2007 Jun;22(5):615-21.
47. Iellamo F, Pigozzi F, Parisi A, Di Salvo V, Vago T, Norbiato G, et al. The stress of competition dissociates neural and cortisol homeostasis in elite athletes. J Sports Med Phys Fitness. 2003 Dec;43(4):539-45.
48. Jedlic B, Hall N, Munroe-Chandler K, Hall C. Coaches' encouragement of athletes' imagery use. Res Q Exerc Sport. 2007 Sep;78(4):351-63.
49. Jurimae J, Maestu J, Purge P, Jurimae T, Soot T. Relations among heavy training stress, mood state, and performance for male junior rowers. Percept Mot Skills. 2002 Oct;95(2):520-6.
50. Kavussanu M. Motivational predictors of prosocial and antisocial behaviour in football. J Sports Sci. 2006 Jun;24(6):575-88.
51. Kemler DS. Sensitivity to sensoriprocessing, self-discrepancy, and emotional reactivity of collegiate athletes. Percept Mot Skills. 2006 Jun;102(3):747-59.
52. Kjormo O, Halvari H. Two ways related to performance in elite sport: the path of self-confidence and competitive anxiety and the path of group cohesion and group goal-clarity. Percept Mot Skills. 2002 Jun;94(3 Pt 1):950-66.
53. Korobeynikov G, Rossokha G, Koniaeva L, Medvedchuk K, Kulinich I. Psychophysiological diagnostics of functional states in sports medicine. Bratisl Lek Listy. 2006;107(5):205-9.
54. Lamont-Mills A, Christensen SA. Athletic identity and its relationship to sport participation levels. J Sci Med Sport. 2006 Dec;9(6):472-8.
55. Manfredini F, Manfredini R, Carrabre JE, Litmanen H, Zhukovskaja L, Dal Follo D, et al. Competition load and stress in sports: a preliminary study in biathlon. Int J Sports Med. 2002 Jul;23(5):348-52.
56. Mann BJ, Grana WA, Indelicato PA, O'Neill DF, George SZ. A survey of sports medicine physicians regarding psychological issues in patient-athletes. Am J Sports Med. 2007 Dec;35(12):2140-7.
57. Martin GL, Vause T, Schwartzman L. Experimental studies of psychological interventions with athletes in competitions: why so few? Behav Modif. 2005 Jul;29(4):616-41.
58. Martinent G, Ferrand C. A cluster analysis of perfectionism among competitive athletes. Psychol Rep. 2006 Dec;99(3):723-38.
59. McClung M, Collins D. "Because I know it will!": placebo effects of an ergogenic aid on athletic performance. J Sport Exerc Psychol. 2007 Jun;29(3):382-94.
60. McDuff DR, Morse ED, White RK. Professional and collegiate team assistance programs: services and utilization patterns. Clin Sports Med. 2005 Oct;24(4):943-58, x-xi.
61. Mellalieu SD, Neil R, Hanton S. Self-confidence as a mediator of the relationship between competitive anxiety intensity and interpretation. Res Q Exerc Sport. 2006 Jun;77(2):263-70.
62. Miller JR, Adamson GJ, Pink MM, Sweet JC. Comparison of preseason, midseason, and postseason neurocognitive scores in uninjured collegiate football players. Am J Sports Med. 2007 Aug;35(8):1284-8.
63. Millet GP, Groslambert A, Barbier B, Rouillon JD, Candau RB. Modelling the relationships between training, anxiety, and fatigue in elite athletes. Int J Sports Med. 2005 Jul-Aug;26(6):492-8.
64. Moreno FJ, Luis V, Salgado F, Garcia JA, Reina R. Visual behavior and perception of trajectories of moving objects with visual occlusion. Percept Mot Skills. 2005 Aug;101(1):13-20.
65. Morgan WP, Raglin JS, O'Connor PJ. Trait anxiety predicts panic behavior in beginning scuba students. International Journal of Sports Medicine 2004 May;25(4):314-22.
66. Mori S, Ohtani Y, Imanaka K. Reaction times and anticipatory skills of karate athletes. Hum Mov Sci. 2002 Jul;21(2):213-30.
67. Nicholls AR, Polman RC. Coping in sport: A systematic review. J Sports Sci. 2007 Jan 1;25(1):11-31.
68. Niess AM, Feherenbach E, Roecker K, Lehmann R, Opavsky L, Dickhuth HH. Individual differences in self-reported heat tolerance. Is there a link to the cardiocirculatory, thermoregulatory and hormonal response to endurance exercise in heat? J Sports Med Phys Fitness. 2003 Sep;43(3):386-92.
69. O'Brien M, Hanton S, Mellalieu SD. Intensity and direction of competitive anxiety as a function of goal attainment expectation and competition goal generation. J Sci Med Sport. 2005 Dec;8(4):423-32.
70. Pearce PZ. A practical approach to the overtraining syndrome. Curr Sports Med Rep. 2002 Jun;1(3):179-83.

71. Pensgaard AM, Roberts GC. Elite athletes' experiences of the motivational climate: the coach matters. Scand J Med Sci Sports. 2002 Feb;12(1):54-9.
72. Perna FM, Antoni MH, Baum A, Gordon P, Schneiderman N. Cognitive behavioral stress management effects on injury and illness among competitive athletes: a randomized clinical trial. Ann Behav Med. 2003 Winter;25(1):66-73.
73. Perseghin G, De Cobelli F, Esposito A, Lattuada G, Terruzzi I, La Torre A, et al. Effect of the sporting discipline on the right and left ventricular morphology and function of elite male track runners: a magnetic resonance imaging and phosphorus 31 spectroscopy study. Am Heart J. 2007 Nov;154(5):937-42.
74. Philippe RA, Seiler R. Sex differences on use of associative and dissociative cognitive strategies among male and female athletes. Percept Mot Skills. 2005 Oct;101(2):440-4.
75. Pierce EF, Jr. Relationship between training volume and mood states in competitive swimmers during a 24-week season. Percept Mot Skills. 2002 Jun;94(3 Pt 1):1009-12.
76. Powell DH. Treating individuals with debilitating performance anxiety: An introduction. J Clin Psychol. 2004 Aug;60(8):801-8.
77. Purge P, Jurimae J, Jurimae T. Changes in recovery-stress state and performance in elite rowers during preparation for major competitions. Percept Mot Skills. 2005 Oct;101(2):375-81.
78. Raglin JS. Psychological factors in sport performance: the Mental Health Model revisited. Sports Med. 2001;31(12):875-90.
79. Raglin JS. The psychology of the marathoner : of one mind and many. Sports Med. 2007;37(4-5):404-7.
80. Robazza B, Bertollo M, Bortoli L. Frequency and direction of competitive anger in contact sports. J Sports Med Phys Fitness. 2006 Sep;46(3):501-8.
81. Rogers NL, Dinges DF. Caffeine: implications for alertness in athletes. Clin Sports Med. 2005 Apr;24(2):e1-13, x-xi.
82. Rouveix M, Duclos M, Gouarne C, Beauvieux MC, Filaire E. The 24 h urinary cortisol/cortisone ratio and epinephrine/norepinephrine ratio for monitoring training in young female tennis players. Int J Sports Med. 2006 Nov;27(11):856-63.
83. Ryan JP, Atkinson TM, Dunham KT. Sports-related and gender differences on neuropsychological measures of frontal lobe functioning. Clin J Sport Med. 2004 Jan;14(1):18-24.
84. Sagar SS, Lavallee D, Spray CM. Why young elite athletes fear failure: consequences of failure. J Sports Sci. 2007 Sep;25(11):1171-84.
85. Sands WA, McNeal JR, Stone MH. Plaudits and pitfalls in studying elite athletes. Percept Mot Skills. 2005 Feb;100(1):22-4.
86. Settles IH, Sellers RM, Alphonse D, Jr. One role or two? The function of psychological separation in role conflict. J Appl Psychol. 2002 Jun;87(3):574-82.
87. Sforza C, Grassi GP, Turci M, Fragnito N, Pizzini G, Ferrario VF. Influence of training on maintenance of equilibrium on a tilting platform. Percept Mot Skills. 2003 Feb;96(1):127-36.
88. Shannon AL, Ferrari R, Russell AS. Alberta rodeo athletes do not develop the chronic whiplash syndrome. J Rheumatol. 2006 May;33(5):975-7.
89. Sherman DK, Kinias Z, Major B, Kim HS, Prenovost M. The group as a resource: reducing biased attributions for group success and failure via group affirmation. Pers Soc Psychol Bull. 2007 Aug;33(8):1100-12.
90. Shields DL, LaVoi NM, Bredemeier BL, Power FC. Predictors of poor sportspersonship in youth sports: personal attitudes and social influences. J Sport Exerc Psychol. 2007 Dec;29(6):747-62.
91. Smith A, Ntoumanis N, Duda J. Goal striving, goal attainment, and well-being: adapting and testing the self-concordance model in sport. J Sport Exerc Psychol. 2007 Dec;29(6):763-82.
92. Smith AL, Balaguer I, Duda JL. Goal orientation profile differences on perceived motivational climate, perceived peer relationships, and motivation-related responses of youth athletes. J Sports Sci. 2006 Dec;24(12):1315-27.
93. St Clair Gibson A, Lambert EV, Rauch LH, Tucker R, Baden DA, Foster C, et al. The role of information processing between the brain and peripheral physiological systems in pacing and perception of effort. Sports Med. 2006;36(8):705-22.
94. Storch EA, Kovacs AH, Roberti JW, Bailey LM, Bravata EA, Storch JB. Strength of religious faith and psychological adjustment in intercollegiate athletes. Psychol Rep. 2004 Feb;94(1):48-50.
95. Tod D, Iredale F, Gill N. 'Psyching-up' and muscular force production. Sports Med. 2003;33(1):47-58.
96. Uphill MA, Jones MV. Antecedents of emotions in elite athletes: a cognitive motivational relational theory perspective. Res Q Exerc Sport. 2007 Mar;78(2):79-89.
97. Vickers JN, Williams AM. Performing under pressure: the effects of physiological arousal, cognitive anxiety, and gaze control in biathlon. J Mot Behav. 2007 Sep;39(5):381-94.
98. Walker N, Thatcher J, Lavallee D. Psychological responses to injury in competitive sport: a critical review. J R Soc Health. 2007 Jul;127(4):174-80.
99. Wang J, Marchant D, Morris T, Gibbs P. Self-consciousness and trait anxiety as predictors of choking in sport. J Sci Med Sport. 2004 Jun;7(2):174-85.
100. Wanner B, Ladouceur R, Auclair AV, Vitaro F. Flow and dissociation: examination of mean levels, cross-links, and links to emotional well-being across sports and recreational and pathological gambling. J Gambl Stud. 2006 Fall;22(3):289-304.
101. Waterhouse J, Atkinson G, Edwards B, Reilly T. The role of a short post-lunch nap in improving cognitive, motor, and sprint performance in participants with partial sleep deprivation. J Sports Sci. 2007 Dec;25(14):1557-66.
102. Wiggins MS, Cremades JG, Lai C, Lee J, Erdmann JB. Multidimensional comparison of anxiety direction and burnout over time. Percept Mot Skills. 2006 Jun;102(3):788-90.
103. Williams SJ, Kendall LR. A profile of sports science research (1983-2003). J Sci Med Sport. 2007 Aug;10(4):193-200.
104. Zahariadis P, Tsorbatzoudis H, Alexandris K. Self-determination in sport commitment. Percept Mot Skills. 2006 Apr;102(2):405-20.
105. Zhang J, Watanabe K. Differences in saccadic latency and express saccades between skilled and novice ball players in tracking predictable and unpredictable targets at two visual angles. Percept Mot Skills. 2005 Jun;100(3 Pt 2):1127-36.

Older Studies or Reviews Recommended by Committee Members

106. Morgan WP. Affective beneficence of vigorous physical activity. Med Sci Sports Exerc. 1985; 17(1):94-100.
107. Morgan WP. Psychogenic factors and exercise metabolism: a review. Med Sci Sports Exerc. 1985 Jun;17(3):309-16.
108. O'Connor PJ and Morgan WP. Athletic performance following rapid traversal of multiple time zones. A review. Sports Medicine. 1990; 10(1): 20-30.
109. Raglin JS, Morgan WP. Influence of exercise and quiet rest on state anxiety and blood pressure. Med Sci Sports Exerc. 1987 Oct;19(5):456-63.

TRAINING WITH MENTAL IMAGERY

110. Beauchamp MR, Bray SR, Albinson JG. Pre-competition imagery, self-efficacy and performance in collegiate golfers. J Sports Sci. 2002 Sep;20(9):697-705.
111. Calabrese P, Messonnier L, Bijaoui E, Eberhard A, Benchetrit G. Cardioventilatory changes induced by mentally imaged rowing. Eur J Appl Physiol. 2004 Mar;91(2-3):160-6.

112. Cumming J, Hall C. Deliberate imagery practice: the development of imagery skills in competitive athletes. J Sports Sci. 2002 Feb;20(2):137-45.
113. Cumming J, Hall C, Starkes JL. Deliberate imagery practice: the reliability of using a retrospective recall methodology. Res Q Exerc Sport. 2005 Sep;76(3):306-14.
114. Fontani G, Migliorini S, Benocci R, Facchini A, Casini M, Corradeschi F. Effect of mental imagery on the development of skilled motor actions. Percept Mot Skills. 2007 Dec;105(3 Pt 1):803-26.
115. Gregg M, Hall C. Measurement of motivational imagery abilities in sport. J Sports Sci. 2006 Sep;24(9):961-71.
116. Gualberto Cremades J. The effects of imagery perspective as a function of skill level on alpha activity. Int J Psychophysiol. 2002 Mar;43(3):261-71.
117. Hardy J, Hall CR, Alexander MR. Exploring self-talk and affective states in sport. J Sports Sci. 2001 Jul;19(7):469-75.
118. Louis M, Guillot A, Maton S, Doyon J, Collet C. Effect of imagined movement speed on subsequent motor performance. J Mot Behav. 2008 Mar;40(2):117-32.
119. MacIntyre T, Moran A, Jennings DJ. Is controllability of imagery related to canoe-slalom performance? Percept Mot Skills. 2002 Jun;94(3 Pt 2):1245-50.
120. Munroe-Chandler KJ, Hall CR, Fishburne GJ, Strachan L. Where, when, and why young athletes use imagery: an examination of developmental differences. Res Q Exerc Sport. 2007 Mar;78(2):103-16.
121. Oishi K, Maeshima T. Autonomic nervous system activities during motor imagery in elite athletes. J Clin Neurophysiol. 2004 May-Jun;21(3):170-9.
122. Rogers RG. Mental practice and acquisition of motor skills: examples from sports training and surgical education. Obstet Gynecol Clin North Am. 2006 Jun;33(2):297-304, ix.
123. Short SE, Tenute A, Feltz DL. Imagery use in sport: mediational effects for efficacy. J Sports Sci. 2005 Sep;23(9):951-60.

EFFECTS OF CONCUSSION

124. Bailey CM, Echemendia RJ, Arnett PA. The impact of motivation on neuropsychological performance in sports-related mild traumatic brain injury. J Int Neuropsychol Soc. 2006 Jul;12(4):475-84.
125. Barr WB, McCrea M. Sensitivity and specificity of standardized neurocognitive testing immediately following sports concussion. J Int Neuropsychol Soc. 2001 Sep;7(6):693-702.
126. Bauman J. Returning to play: the mind does matter. Clin J Sport Med. 2005 Nov;15(6):432-5.
127. Broglio SP, Macciocchi SN, Ferrara MS. Sensitivity of the concussion assessment battery. Neurosurgery. 2007 Jun;60(6):1050-7; discussion 7-8.
128. Brooks DA. Use of computer based testing of youth hockey players with concussions. NeuroRehabilitation. 2007;22(3):169-79.
129. Chen JK, Johnston KM, Petrides M, Ptito A. Neural substrates of symptoms of depression following concussion in male athletes with persisting postconcussion symptoms. Arch Gen Psychiatry. 2008 Jan;65(1):81-9.
130. Collie A, Makdissi M, Maruff P, Bennell K, McCrory P. Cognition in the days following concussion: comparison of symptomatic versus asymptomatic athletes. J Neurol Neurosurg Psychiatry. 2006 Feb;77(2):241-5.
131. Collie A, McCrory P, Makdissi M. Does history of concussion affect current cognitive status? Br J Sports Med. 2006 Jun;40(6):550-1.
132. Collins MW, Lovell MR, Iverson GL, Cantu RC, Maroon JC, Field M. Cumulative effects of concussion in high school athletes. Neurosurgery. 2002 Nov;51(5):1175-9; discussion 80-1.
133. De Beaumont L, Brisson B, Lassonde M, Jolicoeur P. Long-term electrophysiological changes in athletes with a history of multiple concussions. Brain Inj. 2007 Jun;21(6):631-44.
134. Gosselin N, Theriault M, Leclerc S, Montplaisir J, Lassonde M. Neurophysiological anomalies in symptomatic and asymptomatic concussed athletes. Neurosurgery. 2006 Jun;58(6):1151-61; discussion -61.
135. Hinton-Bayre AD, Geffen G. Severity of sports-related concussion and neuropsychological test performance. Neurology. 2002 Oct 8;59(7):1068-70.
136. Iverson GL, Gaetz M, Lovell MR, Collins MW. Relation between subjective fogginess and neuropsychological testing following concussion. J Int Neuropsychol Soc. 2004 Oct;10(6):904-6.
137. Killam C, Cautin RL, Santucci AC. Assessing the enduring residual neuropsychological effects of head trauma in college athletes who participate in contact sports. Arch Clin Neuropsychol. 2005 Jul;20(5):599-611.
138. Lovell MR, Iverson GL, Collins MW, Podell K, Johnston KM, Pardini D, et al. Measurement of symptoms following sports-related concussion: reliability and normative data for the post-concussion scale. Appl Neuropsychol. 2006;13(3):166-74.
139. Pellman EJ, Lovell MR, Viano DC, Casson IR. Concussion in professional football: recovery of NFL and high school athletes assessed by computerized neuropsychological testing--Part 12. Neurosurgery. 2006 Feb;58(2):263-74; discussion -74.
140. Piland SG, Motl RW, Guskiewicz KM, McCrea M, Ferrara MS. Structural validity of a self-report concussion-related symptom scale. Med Sci Sports Exerc. 2006 Jan;38(1):27-32.
141. Podlog L, Eklund RC. Assisting injured athletes with the return to sport transition. Clin J Sport Med. 2004 Sep;14(5):257-9.
142. Shapcott EJ, Bloom GA, Johnston KM, Loughead TM, Delaney JS. The effects of explanatory style on concussion outcomes in sport. NeuroRehabilitation. 2007;22(3):161-7.
143. Register-Mihalik J, Guskiewicz KM, Mann JD, Shields EW. The effects of headache on clinical measures of neurocognitive function. Clin J Sport Med. 2007 Jul;17(4):282-8.

TRAINING AND PERFORMANCE ISSUES FOR FEMALE ATHLETES

144. Aizawa K, Nakahori C, Akimoto T, Kimura F, Hayashi K, Kono I, et al. Changes of pituitary, adrenal and gonadal hormones during competition among female soccer players. J Sports Med Phys Fitness. 2006 Jun;46(2):322-7.
145. Brooks JE, Bull SJ. Derogation of student female athletes who consult a sport psychologist: an alternative perspective on the negative halo effect. J Sci Med Sport. 2001 Mar;4(1):39-47.
146. Forsberg S, Lock J. The relationship between perfectionism, eating disorders and athletes: a review. Minerva Pediatr. 2006 Dec;58(6):525-36.
147. Lal S, Hoch AZ. Factors that affect the young female athlete. Phys Med Rehabil Clin N Am. 2007 Aug;18(3):361-83, vii.
148. Prather H, Hunt D. Issues unique to the female runner. Phys Med Rehabil Clin N Am. 2005 Aug;16(3):691-709.
149. Wiggins MS, Lai C, Deiters JA. Anxiety and burnout in female collegiate ice hockey and soccer athletes. Percept Mot Skills. 2005 Oct;101(2):519-24.
150. Zamboanga BL, Horton NJ, Leitkowski LK, Wang SC. Do good things come to those who drink? A longitudinal investigation of drinking expectancies and hazardous alcohol use in female college athletes. J Adolesc Health. 2006 Aug;39(2):229-36.
151. Bruening JE, Armstrong KL, Pastore DL. Listening to the voices: the experiences of African American female student athletes. Res Q Exerc Sport. 2005 Mar;76(1):82-100.

RISK-TAKING BEHAVIOR AND ATHLETIC PERFORMANCE

152. Ford JA. Substance use among college athletes: a comparison based on sport/team affiliation. J Am Coll Health. 2007 May-Jun;55(6):367-73.
153. Huang JH, Jacobs DF, Derevensky JL, Gupta R, Paskus TS. A national study on gambling among US college student-athletes. J Am Coll Health. 2007 Sep-Oct;56(2):93-9.
154. Huang JH, Jacobs DF, Derevensky JL, Gupta R, Paskus TS. Gambling and health risk behaviors among U.S. college student-athletes: findings from a national study. J Adolesc Health. 2007 May;40(5):390-7.
155. Miller TW, Adams JM, Kraus RF, Clayton R, Miller JM, Anderson J, et al. Gambling as an addictive disorder among athletes: clinical issues in sports medicine. Sports Med. 2001;31(3):145-52.
156. Weinstock J, Whelan JP, Meyers AW, Watson JM. Gambling behavior of student-athletes and a student cohort: what are the odds? J Gambl Stud. 2007 Mar;23(1):13-24.
157. Wetherill RR, Fromme K. Alcohol use, sexual activity, and perceived risk in high school athletes and non-athletes. J Adolesc Health. 2007 Sep;41(3):294-301.

RELATIONSHIP BETWEEN LIFESTYLE AND IMMUNE FUNCTION

158. Clow A, Hucklebridge F. The impact of psychological stress on immune function in the athletic population. Exerc Immunol Rev. 2001;7:5-17.
159. Gleeson M. Immune function in sport and exercise. J Appl Physiol. 2007 Aug;103(2):693-9.
160. Perna FM, Antoni MH, Baum A, Gordon P, Schneiderman N. Cognitive behavioral stress management effects on injury and illness among competitive athletes: a randomized clinical trial. Ann Behav Med. 2003 Winter;25(1):66-73.

OTHER TOPICS

161. Bishop DT, Karageorghis CI, Loizou G. A grounded theory of young tennis players use of music to manipulate emotional state. J Sport Exerc Psychol. 2007 Oct;29(5):584-607.
162. Desbrow B, Leveritt M. Well-trained endurance athletes' knowledge, insight, and experience of caffeine use. Int J Sport Nutr Exerc Metab. 2007 Aug;17(4):328-39.

Appendix D

Research on Managing Information Overload in Soldiers Under Stress

This appendix provides background information on research in augmented cognition (AugCog) conducted by the Defense Advanced Research Projects Agency (DARPA) and the U.S. Army and discusses several of the neuroscience research and engineering challenges.

THE DARPA AUGCOG PROGRAM

DARPA began its research program in the area of augmented cognition in 2001 with a focus on challenges and opportunities presented by the real-time monitoring of cognitive state with physiological sensors (Augmented Cognition International Society, 2008a). The program, now known as the Improving Warfighter Information Intake Under Stress Program, was initially called the Augmented Cognition program, or AugCog. At first it had two stated purposes: (1) to gain battlefield information superiority and (2) neurology-related clinical applications. Linked to the idea of network-centric warfare, the stated rationale for AugCog was that "order of magnitude increases in available, net thinking power resulting from linked human-machine dyads will provide such clear information superiority that few rational individuals or organizations would challenge under the consequences of mortality" (Schmorrow and McBride, 2004). Another early AugCog objective was stated as the enhancement of a single operator's capabilities such that he or she could carry out the functions of three or more individuals (McDowell, 2002).

It was proposed that information superiority could be achieved by enhancing an operator's cognitive abilities and by measuring neurological and physiologic variables. One aspect of the enhancement would involve tracking markers of cognitive state, such as respiration, heart rate, or eye movement, using devices for electroencephalography (EEG) or functional optical imaging. The operator would wear a headset that contained such devices and might be connected to other devices as well—perhaps galvanic skin response sensors or pressure sensors in a seat—interacting with them by means of a traditional computer interface device such as a mouse or a joystick. Another aspect of enhancing cognitive ability would be dynamic control of the rate as well as the source of information by a "communications scheduler." The former DARPA program manager envisioned that technologies developed by the program would be operational within 10 years. He predicted that within 20 years, the technology would be woven into the fabric of our daily lives (Augmented Cognition International Society, 2008b).

Proof-of-concept work for AugCog occurred in two phases. In Phase 1, researchers attempted to detect changes in cognitive activity in near real time in an operationally relevant setting. One relatively large demonstration, called the Technical Integration Experiment, was conducted with mixed results (St. John et al., 2003). Its objective was to determine which psychophysiological measurements could consistently detect changes in cognitive activity during a supervisory control task. Using 20 gauges of cognitive state (CSGs) such as EEG, functional near infrared (fNIR), and body posture, as well as input device measures (mouse pressure and mouse clicks), eight participants completed a series of four simplified aircraft monitoring and threat response tasks in 1 hour. Eleven of the CSGs, including fNIR and EEG measures, were reported to be significant and reliable for at least one independent variable, but no CSG was significant and reliable across all of three independent variables (St. John et al., 2004). Two CSGs, mouse clicks and mouse pressure, demonstrated statistically significant results across two of the independent variables, but these were probably already highly correlated.

There were significant problems with technical integration, several of which were acknowledged by St. John and his colleagues in *Technical Integration Experiment* (2004). These included questions about the study design constructs and their external validity, the statistical methods applied to the data, important data missing from the report of results, and the familiar problem in psychophysiological research of noisy data (substantial variability across test participants, test runs, etc.). Given the number and severity of confounding factors, the results of the study are preliminary at best and in no way constitute unequivocal scientific evidence that the CSGs identified as statistically significant can effectively detect change in cognitive activity in a complex human supervisory control task.

A second set of four experiments was conducted in Phase 2 of AugCog. The stated objective of Phase 2 was to manipulate an operator's cognitive state as a result of near-real-time psychophysiological measurements (Dorneich et al., 2005). The experiments used a video game environment to simulate military operations in urban terrain (MOUT) in either a desktop setting or a motion-capture laboratory. In addition to the primary task, navigating through the MOUT, participants had to distinguish friends from foes while monitoring and responding to communications. A communications scheduler, part of the Honeywell Joint Human–Automation Augmented Cognition System,[1] determined operator workload via a cognitive state profile (CSP) and prioritized incoming messages accordingly. The CSP was an amalgam of signals from cardiac interbeat interval, heart rate, pupil diameter, EEG P300, cardiac quasi-random signal (QRS), and EEG power at the frontal (FCZ) and central midline (CPZ) sites (Dorneich et al., 2005).

As in the Phase 1 experiment, there were only a few participants (16 or fewer) in each of the four Phase 2 experiments. Construct validity and statistical models were questionable, with significant experimental confounds. There is no open account of how the neurological and physiological variables were combined to form the CSP, making independent peer-researched replication of these experiments difficult. In light of these concerns, claims such as a 100 percent improvement in message comprehension, a 125 percent improvement in situation awareness, a 150 percent increase in working memory, and a more than 350 percent improvement in survivability should be considered tentative. In addition, the authors claim anecdotal evidence that their CSGs can indicate operator inability to comprehend a message (Dorneich et al., 2005).

The focus in these experiments appears to have been on generating measurable outcomes on a very tight time schedule. Most of the technical data on the performance of the actual sensors and of the signal processing and combination algorithms were not published. This information would have been useful for further scientific evaluation and confirmation of the reported results.

The problem with AugCog as a development lies less in the intrinsic concept of managing cognitive workload through neural and physiological feedback to a smart information system than in the assumptions that were made about the maturity of the technologies required to implement such a system and thus about the time frame for an initial operational capability. Unfortunately, no follow-on studies have reported how the successful CSGs could or would be combined in an operational system. The engineering obstacles to combining EEG, fNIR, and eye-tracking devices are substantial. Unless dramatic leaps are made soon in the miniaturization of these technologies and in improved signal-processing algorithms, the realization of a single headset that can combine all—or even a subset—of these technologies is at least a decade away.

Other engineering problems, such as how to measure EEG signals in a dynamic, noisy environment, have not been addressed, at least in the open literature. Basic sensor system engineering problems like these will be critical to any operational deployment of these technologies. A similar engineering problem underlay the use of the eye-tracking devices assumed for AugCog applications. These devices currently require a sophisticated head-tracking device in addition to the eye-tracking device, and encapsulating this technology into an unobtrusive device that can be worn in the field appears also to be at least 10 years in the future.

In addition to hardware limitations on the use of neural and physiological technologies in an operational field setting, the software/hardware suite required to interpret cognitive state reliably in real time is beyond current capabilities, particularly in the highly dynamic, stochastic settings typical of command-and-control environments. The experiments for the AugCog program were conducted under controlled laboratory conditions. While this is to be expected for preliminary, proof-of-concept studies, such a limitation constrains the extrapolation of the reported results. For example, the communications scheduler in the Phase 2 experiments made changes in information presentation based on gross differences in perceived cognitive state. In actual battlefield conditions, the amount of task-relevant information and the degrees of freedom in cognitive state will require more precision and reliability in ascertaining an operator's condition and making situation-appropriate adjustments rather than limiting access, perhaps inappropriately, to information that may be critical for a real-time decision. Not only must the sensors and signal-processing algorithms improve substantially; significant advances are also needed in decision-theoretic modeling. In particular, these models will have to accommodate a significant range of individual variability.

Overall, the AugCog goal of enhancing operator performance through psychophysiological sensing and automation-based reasoning is desirable but faces major challenges as an active information filter. Suppose a system is implemented that can change information streams and decrease the volume by filtering incoming information presented to a user. How is the system to know that its filtering in a specific situation is both helpful to this user and passes along the correct information for the current situation? The system software must correctly determine an optimal cognitive load for an individual in a dynamic, highly uncertain context and decide

[1]Honeywell was the prime contractor for the DARPA AugCog program and remains the prime contractor for the follow-on Army program in augmented cognition.

which information to emphasize and which information to minimize or even filter out altogether. The problem is that, in command-and-control settings, there are no general principles for what information is truly optimal. Before deploying such an active-filter system, which controls inputs to a military operator, rigorous testing and validation would have to demonstrate that at the least, it does no worse than an unaided soldier.

ARMY FOLLOW-ON WORK

When the DARPA AugCog program formally ended, the Army continued working with portions of the concept at the U.S. Army Natick Soldier Research, Development and Engineering Center. The original goal of the Army effort was to incorporate AugCog technology into the Army Future Force Warrior Program by 2007 (U.S. Army, 2005). The primary focus of the effort has since changed from operational to training applications; moreover, the technology focus has narrowed to the use of EEG and electrocardiography sensors instead of the array of neurological and physiological sensors envisioned for the DARPA AugCog scheme (Boland, 2008).

An experiment was conducted to extend the previous DARPA effort, directed by the same Honeywell team that performed the set of experiments described above. This Army-sponsored experiment focused on developing an experimental test bed for mobile cognitive-state classification and testing it in a dismounted-soldier field setting using the previously discussed communications scheduler (Dorneich et al., 2007). The authors developed an EEG headset connected to a laptop computer worn in a backpack by the test subject. This laptop supported the signal processing algorithms, the communications scheduler, and other experimental testing elements. For the experiment, eight subjects with no military experience completed a 1-hour navigation and communication task with a handheld radio, a personal digital assistant, and a 35-lb backpack. The authors reported that, with the communications scheduler prioritizing messages based on whether the subjects were in a low-task-load or high-task-load condition, mission performance metrics improved from 68 percent to 96 percent with cognitive-state mitigation.

As with the earlier experiments, significant confounds limited the validity of the results. Problems were reported with movement-induced signal noise, as well as significant loss of data that reduced the subject pool to just four individuals for a portion of the experiment. In addition, the approach to classifying cognitive state was extremely limited in state estimation (i.e., costs of actions were not considered), and it depended on relatively short temporal gaps between training and testing (Dorneich et al., 2007). This latter constraint in an operational setting means that soldiers would require extended training to "condition" the system before each mission. Since actual combat never follows a carefully planned script, an issue yet to be addressed is how a priori classification training can ever be based on events that are real enough to give reliable results. The ultimate problem is not that the information filter might fail to accurately gauge the cognitive state of the user, but that it might act in a way that results in a bad decision.

REFERENCES

Augmented Cognition International Society. 2008a. History: Emergence of augmented cognition. Available at http://www.augmentedcognition.com/history.htm. Last accessed August 17, 2008.

Augmented Cognition International Society. 2008b. Frequently Asked Questions. Available at http://www.augmentedcognition.org/faq.htm#q2. Last accessed August 17, 2008.

Boland, R. 2008. Army uses advanced systems to understand what soldiers know. Signal. Fairfax, Va.: Armed Forces Communications and Electronics Association.

Dorneich, M.C., P.M. Ververs, M. Santosh, and S.D. Whitlow. 2005. A joint human–automation cognitive system to support rapid decision-making in hostile environments. Pp. 2390- 2395 in IEEE International Conference on Systems, Man and Cybernetics, Vol. 3. Los Alamitos, Calif.: IEEE Publications Office.

Dorneich, M.C., S.D. Whitlow, S. Mathan, P.M. Ververs, D. Erodogmus, A. Adami, M. Pavel, and T. Lan. 2007. Supporting real-time cognitive state classification on a mobile system. Journal of Cognitive Engineering and Decision Making 1(3): 240-270.

McDowell, P. 2002. The MOVES Institute's Context Machine Project. Available online at http://www.movesinstitute.org/~mcdowell/augCog/. Last accessed May 12, 2008.

Schmorrow, D., and D. McBride. 2004. Introduction: Special issue on augmented cognition. International Journal of Human-Computer Interaction 17(2): 127-130.

St. John, M., D.A. Kobus, and J.G. Morrison. 2003. DARPA Augmented Cognition Technical Integration Experiment (TIED). Technical Report 1905, December. San Diego, Calif.: U.S. Navy Space and Naval Warfare Systems Center.

St. John, M., D.A. Kobus, J.G. Morrison, and D. Schmorrow. 2004. Overview of the DARPA Augmented Cognition technical integration experiment. International Journal of Human–Computer Interaction 17(2), 131-149.

U.S. Army. 2005. Mission: Augmented cognition. The Warrior. Natick, Mass.: Public Affairs Office, U.S. Army Natick Soldier Research, Development and Engineering Center.